Mansfield Merriman

Elements of precise surveying and geodesy

Mansfield Merriman

Elements of precise surveying and geodesy

ISBN/EAN: 9783337275471

Printed in Europe, USA, Canada, Australia, Japan

Cover: Foto ©Andreas Hilbeck / pixelio.de

More available books at **www.hansebooks.com**

ELEMENTS

OF

PRECISE SURVEYING

AND

GEODESY.

BY

MANSFIELD MERRIMAN,
PROFESSOR OF CIVIL ENGINEERING IN LEHIGH UNIVERSITY.

FIRST EDITION.

FIRST THOUSAND.

NEW YORK:
JOHN WILEY & SONS.
LONDON: CHAPMAN & HALL, LIMITED.
1899.

ROBERT DRUMMOND, PRINTER, NEW YORK.

CONTENTS.

CHAPTER I.

THE METHOD OF LEAST SQUARES.

CHAPTER II.

PRECISE PLANE TRIANGULATION.

Chapter III.

BASE LINES.

Chapter IV.

LEVELING.

Chapter V.

ASTRONOMICAL WORK.

Chapter VI.

SPHERICAL GEODESY.

Chapter VII.

SPHEROIDAL GEODESY.

Chapter VIII.

GEODETIC COORDINATES AND PROJECTIONS.

Chapter IX.

GEODETIC TRIANGULATION.

Chapter X.

THE FIGURE OF THE EARTH.

Chapter XI.

TABLES.

PRECISE SURVEYING AND GEODESY.

CHAPTER I.

THE METHOD OF LEAST SQUARES.

1. ERRORS OF OBSERVATIONS.

The Method of Least Squares furnishes processes of computation by which the most probable values of quantities are found from the results of measurements. The simplest case is that of a quantity which is directly measured several times with equal precision; here it is universally agreed that the arithmetic mean of the several values is the most probable value of the quantity.

When a quantity is measured the result of the operation is a numerical value called an observation. If Z be the true value of a quantity and M_1 and M_2 be two observations upon it, then $Z - M_1$ and $Z - M_2$ are the errors of those observations.

Constant or systematic errors are those which result from causes well understood and which can be computed or eliminated. As such may be mentioned: theoretical errors, like the effects of refraction upon a vertical angle, or the effects of temperature upon a steel tape, which can be computed when proper data are known and hence need not be classed as real errors; instrumental errors, like the effects of an imperfect adjustment of an instrument, which can be removed by taking proper precautions in advance; and personal errors which are due to the habits of the observer, who may, for example, always give the reading of a scale too great. All these causes are to be carefully investigated and the resultant errors removed from the final observations.

Mistakes are errors due to such serious mental confusion that the observation cannot be regarded with any confidence, as for instance writing 53 when 35 is intended. Observations affected with mistakes must be rejected, although when these are of small magnitude it is sometimes not easy to distinguish them from errors.

Accidental errors are those that still remain after all constant errors and mistakes have been carefully investigated and eliminated. Such, for example, are the errors in leveling arising from sudden expansions of the bubble and standards, or from the effects of the wind, or from irregular refraction. They also arise from the imperfections of human touch and sight, which render it difficult to handle instruments delicately or to read verniers with perfect accuracy. These are the errors that exist in the final observations and whose discussion forms the subject of this chapter.

However carefully the measurements be made, the final observations do not agree; all of these observations cannot be correct, since the quantity has only one value, and each of them can be regarded only as an approximation to the truth. The absolutely true value of the quantity in question cannot be ascertained, but instead of it one must be determined, derived from the combination of the observations, which shall be the " most probable value," that is to say, the value which is probably most nearly to the truth.

The difference between the most probable value of a quantity and an observation is called the ' residual error ' of that observation. Thus, if z be the most probable value of a quantity derived from the observations M_1 and M_2, and v_1 and v_2 be the residual errors, then

$$v_1 = z - M_1, \quad v_2 = z - M_2. \qquad (1)$$

When the measurements are numerous and precise the most probable value z does not greatly differ from the true value Z, and the residuals do not greatly differ from the true errors.

Prob. 1. Eight measurements of a line give the values 186.4, 186.3, 186.2, 186.3, 186.3, 186.2, 185.9 and 186.4 inches, and its most probable length is their arithmetic mean. Compute the eight residual errors; find the sum of the positive residuals and the sum of the negative residuals.

2. Law of Probability of Error.

The probability of an error is the ratio of the number of errors of that magnitude to the total number of errors. If there be 100 observations of an angle which give 27 errors lying between 1″ and 2″ the probability that an error lies between these limits is 0.27. Probabilities are thus measured by numbers lying between 0 (impossibility) and 1 (certainty).

A marksman firing at a target with the intention of hitting the center may be compared to an observer, the position of a shot on the target to an observation, and its distance from the center to an error. If the marksman be skilled and all horizontal errors, like the effect of gravity, be eliminated in the sighting of the rifle, it is recognized that the deviations of the shots, or errors, are quite regular and symmetrical. First, it is noticed that small errors are more frequent than large ones; secondly, that errors on one side are about as frequent as on the other; and thirdly, that very large errors do not occur. Moreover, it is known that the greater the skill of the marksman the nearer are his shots to the center of the target.

As an illustration a record of one thousand shots fired

from a battery gun at a target six hundred feet distant may be considered. The target was a rectangle fifty-two feet long by eleven feet high, and the point of aim was its central horizontal line. All the shots struck the target, and the record of the number in the eleven horizontal divisions, each one foot in width, is as follows:

In top division	1 shot
In second division	4 shots
In third division	10 shots
In fourth division	89 shots
In fifth division	190 shots
In middle division	212 shots
In seventh division	204 shots
In eighth division	193 shots
In ninth division	79 shots
In tenth division	16 shots
In bottom division	2 shots
Total	1 000 shots

The figure shows by means of ordinates the distribution of these shots; A being the top division, O the middle, and B the bottom division. It will be observed that there is a slight preponderance of shots below the middle, and there is reason to believe that this is due to a constant error of gravitation not entirely eliminated in the sighting of the gun. If this series of shots were to be repeated again under exactly similar conditions, it might be fair to infer that 0.212 would be the probability of a given shot striking the middle division, and that 0.001 would be the probability of striking the top division. Thus the probability of an error decreases with the magnitude of the error.

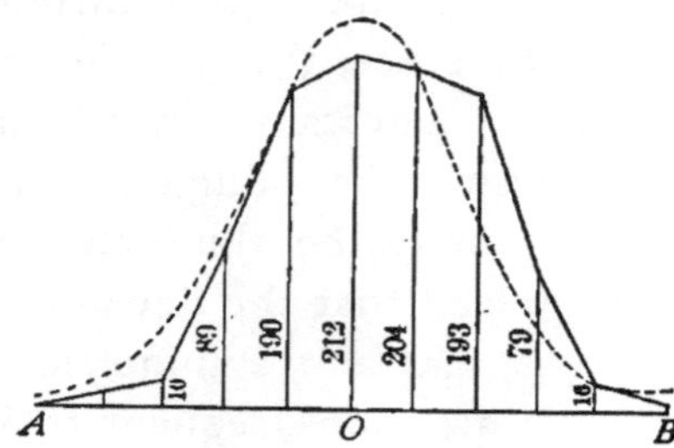

In treatises on the Method of Least Squares the theory of mathematical probability is applied to the deduction of the relation between an error x and its probability y. The equation deduced is

$$y = ce^{-h^2x^2}, \qquad (2)$$

where c and h are constants that depend upon the precision of the measurements and e is the number 2.71828···. This equation expresses the law of probability of accidental errors of observations. It shows that y has its greatest value when

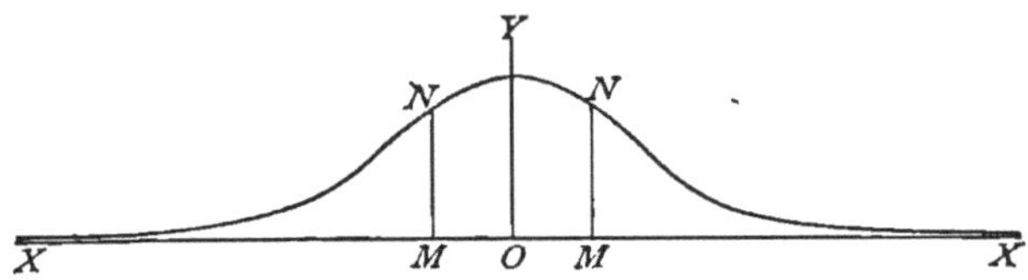

x is zero, that y becomes very small when x is very large, and that the same value of y is given by equal positive and negative values of x. The figure shows the curve expressed by the equation, x and y being parallel to the axes OX and OY, and OM being any error whose probability y is given by the ordinate MN.

This law is deduced under the supposition of a very large number of errors, and hence in a particular case close agreement is not to be expected. For any series of errors the values of c and h^2 can be computed and the theoretic number of errors can then be compared with those actually observed. For example, in the above case of the shots on the target, the value of c can be found to be 0.234 and that of h^2 to be 0.173, and the following comparison shows the agreement of practice with theory:

For division No.	1	2	3	4	5	6	7	8	9	10	11
Actual shots	1	4	10	89	190	212	204	193	79	16	2
Theoretic shots	3	15	50	118	197	234	197	118	50	15	3

The dotted curve on the graphic representation shows the theoretic distribution of the shots. In general it may be said that the results of observation are in good agreement with the theoretic law, and that this agreement is closer the greater the number of errors considered.

Prob. 2. Given the equation $y = 0.234e^{-0.173x^2}$ to compute the values of y when the values of x are 1, 2, 3, 4 and 5.

3. The Principle of Least Squares.

The Method of Least Squares sets forth the processes by which the most probable values of observed quantities are derived from the observations. The foundation of the method is the following principle:

> In observations of equal precision the most probable values of observed quantities are those that render the sum of the squares of the residual errors a minimum.

This principle was first enunciated by Legendre in 1805, and has since been universally accepted and used as the basis of the science of the adjustment of observations. Starting with the law of probability of error enunciated in the last Article it may be proved in the following manner:

Let n observations result from measurements of equal precision, so that c and h in (2) are the same in each case. Then the probability of the error x_1 in the first observation is $ce^{-h^2x_1^2}$, that of the error x_2 is $ce^{-h^2x_2^2}$, and that of the error x_n is $ce^{-h^2x_n^2}$. If the observations be independent of each other, the probability of the simultaneous occurrence of the errors x_1, x_2, . . . x_n is the product of their respective probabilities, or that probability is

$$P = c^n e^{-h^2(x_1^2 + x_2^2 + \ldots + x_n^2)}.$$

Now the true values of the errors x_1, x_2, . . . x_n and those of the quantities observed cannot be found, but the best that can be done is to find their most probable values, namely the values that give the greatest probability P. The greatest value of P occurs when $x_1^2 + x_2^2 + \ldots + x_n^2$ has its least value, that is, the most probable values are those that render the sum of the squares of the errors a minimum. Let v_1, v_2, ... v_n represent the most probable values of the errors x_1, x_2, ... x_n, then

$$v_1^2 + v_2^2 + \ldots + v_n^2 = \text{a minimum} \qquad (3)$$

is an algebraic statement of the fundamental principle of least squares.

An application of this principle to the common case of direct observations on a single quantity will now be given. Let z be the most probable value of the quantity whose n observed values are M_1, M_2, ... M_n, all being of equal precision. Then the residual errors are

$$v_1 = z - M_1, \quad v_2 = z - M_2, \ \ldots \ v_n = z - M_n;$$

and from the fundamental principle (3),

$$(z - M_1)^2 + (z - M_2)^2 + \ldots + (z - M_n)^2 = \text{a minimum}.$$

To find the value of z which renders this expression a minimum it is to be differentiated and the derivative placed equal to zero, giving

$$2(z - M_1) + 2(z - M_2) \ldots + 2(z - M_n) = 0,$$

from which the value of z is found to be

$$z = \frac{M_1 + M_2 + \ldots + M_n}{n},$$

that is, the most probable value of the observed quantity is the arithmetic mean of the observations.

It has been universally recognized from the earliest times that the arithmetic mean gives the most probable value of a quantity which has been measured several times with equal care. Indeed some authors have regarded this as an axiom and used it to deduce the law of probability of error stated in equation (2). It should be noted that the method of the arithmetic mean only applies to equally good observations on a single quantity, and that it cannot be used when observations are made on several related quantities. For instance, let an angle be measured and found to be $60\frac{1}{2}$ degrees, and again be measured in two parts, one being found to be 40 and the other 20 degrees. The proper adjusted value of the angle is not, as might at first be supposed, the mean of $60\frac{1}{2}$ and 60, which is $60\frac{1}{4}$ degrees, but, as will be seen later, it is $60\frac{1}{3}$ degrees, a result which requires each observation to be corrected the same amount.

Prob. 3. Four measurements of a base line give the observations 1472.34 feet, 1471.99 feet, 1472.25 feet, and 1472.14 feet. Compute the sum of the residual errors, and the sum of the squares of the residuals.

4. Weighted Observations.

Observations have equal precision when all the measurements are made with the same care, or when no reason can be assigned to suppose that one is more reliable than another: they are then said to have equal "weight." Weights are numbers expressing the relative practical worth of observations, so that an observation of weight p is worth p times as much as an observation of weight unity. Thus if a line be measured five times with the same care, three measurements giving 950.6 feet and two giving 950.4 feet, then the numbers 3 and 2 are the weights of the observations 950.6 and 950.4. Thus "950.6 with a weight of 3" expresses the same as the number 950.6 written down three times and regarded each time as having a weight of unity; or "950.6 with a weight of 3" might mean that 950.6 is the arithmetic mean of three observations of weight unity.

Let M_1, M_2, . . . M_n be n observations made upon quantities whose most probable values are to be found. Let the residual errors be v_1, v_2, . . . v_n, and by the principle of least squares the values to be found for the quantities must be such that

$$v_1^2 + v_2^2 + \ldots + v_n^2 = \text{a minimum.}$$

Now suppose that there are p_1 observations having the value M_1, or that M_1 has the weight p_1; also that M_2 and M_3 have the weights p_2 and p_3. Then there will be p_1 residuals having the value v_1, p_2 having the value v_2, and p_3 having the value v_3. Thus the condition becomes

$$p_1 v_1^2 + p_2 v_2^2 + \ldots + p_n v_n^2 = \text{a minimum.} \qquad (4)$$

Hence a more general statement of the principle of least squares is:

> In observations of unequal precision the most probable values of the observed quantities are those which render the sum of the weighted squares of the residual errors a minimum.

Here it is seen that the term "weighted square of a residual" means the product of the square of the residual by its weight.

An application of this principle to the case of weighted observations on a single quantity will now be made. Let z be the most probable value of the quantity whose observed values are $M_1, M_2, \ldots M_n$ having the weights $p_1, p_2, \ldots p_n$. Then the residuals are $z - M_1$, $z - M_2$, $\ldots z - M_n$, and from the general principle of least squares given by (3),

$$p_1(z - M_1)^2 + p_2(z - M_2)^2 + \ldots + p_n(z - M_n)^2 = \text{a minimum.}$$

The first derivative of this, placed equal to zero, gives

$$p_1(z - M_1) + p_2(z - M_2) + \ldots + p_n(z - M_n) = 0,$$

from which the most probable value of z is

$$z = \frac{p_1 M_1 + p_2 M_2 + \ldots + p_n M_n}{p_1 + p_2 + \ldots + p_n}.$$

The value of z thus found is sometimes called the weighted arithmetic mean, and the method of computing it is frequently expressed by the rule: Multiply each observation by its weight and divide the sum of the products by the sum of the weights.

Prob. 4. Prove the principle (4) directly from the law of probability of error given by (2), assuming that h^2 represents the weight of the observation whose error is x.

5. Observation Equations.

When observations are taken of several related quantities, the measurements are usually made upon functions of those quantities. Thus the sum and difference of two quantities might be observed instead of the quantities themselves. Such measurements produce "indirect observations" which are generally represented by equations called "observation equations." To illustrate how they arise, let the following practical case be considered. Let O represent a bench-mark, and X, Y, Z, three points whose elevations above O are to be determined. Let five lines of levels be run, giving the results

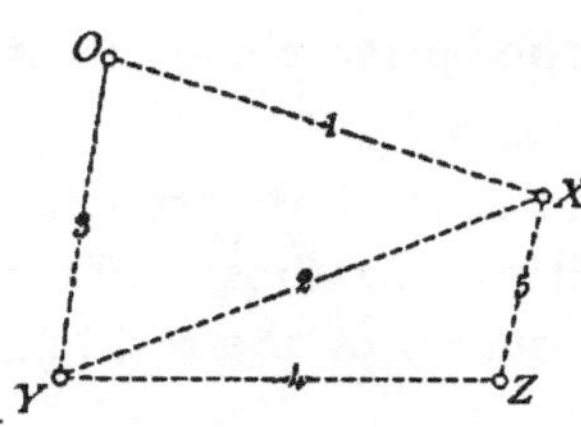

Observation 1. X above $O =$ 10.35 feet.
Observation 2. Y above $X =$ 7.25 feet.
Observation 3. Y above $O =$ 17.63 feet.
Observation 4. Y above $Z =$ 9.10 feet.
Observation 5. Z below $X =$ 1.94 feet.

Here it will be at once perceived that the measurements are discordant; if observations 1 and 2 are taken as correct, the elevation of X is 10.35 feet, and that of Y is 17.60 feet; if 2 and 3 are correct, then X is 10.38 feet and Y is 17.63 feet; and it will be found impossible to deduce values that will exactly satisfy all the observations. Let the elevations of the points X, Y, and Z above O be denoted by x, y, and z, then the observations furnish the following equations:

$$
\begin{aligned}
1. \quad & x = 10.35, \\
2. \quad & y - x = 7.25, \\
3. \quad & y = 17.63, \\
4. \quad & y - z = 9.10, \\
5. \quad & x - z = 1.94.
\end{aligned}
$$

The number of these equations is five, the number of the unknown quantities is three, and hence an exact solution cannot be made. The best that can be done is to find values for x, y, and z which are the most probable, and these will be found in the next Article by the help of the principle of least squares.

Observations are called "direct" when made upon the quantity whose value is sought, and "indirect" when made upon functions of the quantities whose values are required. Thus in the above example the first and third observations are direct, and the others are indirect, being made upon differences of elevation instead of upon the elevations themselves. Indirect observations are of frequent occurrence in the operations of precise surveying.

Quantities are said to be "independent" when each can vary without affecting the value of the others; thus in the above example the elevation of any one station above the bench-mark O is entirely independent of the elevations of the others, or in other words there is no necessary relation between the values of x, y, and z.

Quantities are said to be "conditioned" when they are so related that a change in one necessarily affects the values of the others; thus if the three angles of a plane triangle be called x, y, and z, it is necessary that $x + y + z = 180°$ and the values to be found for the angles must satisfy this condition. In stating observation equations it will often be found best to select the quantities to be determined in such a way that they shall be independent; thus if the three angles of a triangle are observed to be 62° 20′ 43″, 36° 14′ 06″, and 81° 25′ 08″, let x and y denote the most probable values of the first and second angles, then the observation equations are

$$\begin{aligned} x &= 62°\ 20'\ 43'', \\ y &= 36\ \ 14\ \ 06, \\ 180° - x - y &= 81\ \ 25\ \ 08, \end{aligned}$$

the last of which may be written

$$x + y = 98^\circ\ 34'\ 52'',$$

and here x and y are independent quantities. Thus by properly limiting the number of unknown quantities these can always be rendered independent of each other.

As a second example of the statement of observation equations take the following values of the angles measured at North Base, Keweenaw Point, on the United States Lake Survey:

$$\begin{aligned} CNM &= \ \ 55^\circ\ 57'\ 58''.68, \\ MNQ &= \ \ 48\ \ \ 49\ \ \ 13\ \ .64, \\ CNQ &= 104\ \ \ 47\ \ \ 12\ \ .66, \\ QNS &= \ \ 54\ \ \ 38\ \ \ 15\ \ .53, \\ MNS &= 103\ \ \ 27\ \ \ 28\ \ .99. \end{aligned}$$

The object of these observations is to find the values of the four angles around the point N; but if x, y, z, and w represent these angles, then $x + y + z + w = 360^\circ$ and the quantities are conditioned. To make the quantities independent only three unknowns should be taken; thus let $CNM = x$, $MNQ = y$, and $QNS = z$, then the observation equations are

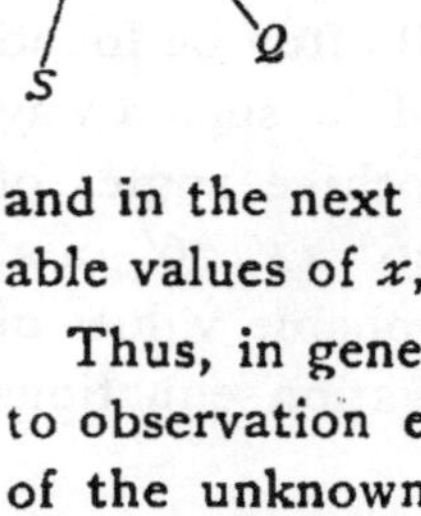

$$\begin{aligned} x &= \ \ 55^\circ\ 57'\ 58''.68, \\ y &= \ \ 48\ \ \ 49\ \ \ 13\ \ .64, \\ x + y &= 104\ \ \ 47\ \ \ 12\ \ .66, \\ z &= \ \ 54\ \ \ 38\ \ \ 15\ \ .53, \\ y + z &= 103\ \ \ 27\ \ \ 28\ \ .99, \end{aligned}$$

and in the next Article it will be shown how the most probable values of x, y, and z are to be found.

Thus, in general, observations upon several quantities lead to observation equations whose number is greater than that of the unknown quantities, and no system of values can be found that will exactly satisfy the observation equations. They may, however, be approximately satisfied by many systems of values; and the problem is to determine that system which is the most probable and hence the best.

Prob. 5. State observation equations for the above example, taking $SNQ = s$, $SNM = t$, $SNC = u$.

6. Indirect Observations of Equal Weight.

When observation equations have been written so that the unknown quantities have no necessary relation to each other, the case is called that of indirect independent observations. Let $M_1, M_2, \ldots M_n$ be n observations of equal weight made upon functions of the unknown quantities x, y, z, etc. Let the observations give the following observation equations:

$$\left.\begin{array}{l} a_1x + b_1y + c_1z + \ldots = M_1, \\ a_2x + b_2y + c_2z + \ldots = M_2, \\ \cdot \quad \cdot \quad \cdot \quad \cdot \quad \cdot \quad \cdot \\ a_nx + b_ny + c_nz + \ldots = M_n, \end{array}\right\} \qquad (5)$$

in which $a_1, a_2, \ldots a_n, b_1, b_2, \ldots b_n$, etc., are known coefficients of the unknown quantities. The most probable values of x, y, z, etc., when found and inserted in the equations will not exactly satisfy them, but leave small residual errors, $v_1, v_2, \ldots v_n$; thus strictly

$$\begin{array}{l} a_1x + b_1y + c_1z + \ldots - M_1 = v_1, \\ a_2x + b_2y + c_2z + \ldots - M_2 = v_2, \\ \cdot \quad \cdot \quad \cdot \quad \cdot \quad \cdot \quad \cdot \quad \cdot \\ a_nx + b_ny + c_nz + \ldots - M_n = v_n, \end{array}$$

and, by the principle of least squares given by (2) in Art. 2, the sum of the squares of these residuals must be a minimum in order to give the most probable values of x, y, and z.

In order to find the condition for the most probable value of x let the terms independent of x in the equations be denoted by $N_1, N_2, \ldots N_n$; then they may be written

$$\begin{array}{l} a_1x + N_1 = v_1, \\ a_2x + N_2 = v_2, \\ \cdot \quad \cdot \quad \cdot \\ a_nx + N_n = v_n. \end{array}$$

Squaring both terms of these equations, and adding, gives

$$(a_1x + N_1)^2 + (a_2x + N_2)^2 + \ldots + (a_nx + N_n)^2 = \Sigma v^2,$$

and this is to be made a minimum to give the most probable value of x. Differentiating it with respect to x, and placing the first derivative equal to zero, there results

$$a_1(a_1x + N_1) + a_2(a_2x + N_2) + \ldots + a_n(a_nx + N_n) = 0, \quad (6)$$

and this is the condition for the most probable value of x. In like manner a similar condition may be stated for each of the other unknown quantities. The conditions thus stated are called "normal equations," and their solution will furnish the most probable values of the required quantities.

The following is hence the rule for the adjustment of observations of equal weight involving several independent quantities:

> For each of the unknown quantities form a normal equation by multiplying each observation equation by the coefficient of that unknown quantity in that equation and adding the results. Then the solution of these normal equations will furnish the most probable values of the unknown quantities.

In forming the normal equations it should be particularly noticed that the signs of coefficients are to be observed in performing the multiplications, and also that when the unknown quantity under consideration does not occur in an observation equation its coefficient is 0.

As an example the five observation equations at the beginning of the last Article will be taken. They may be written

$$\begin{array}{lrrrl}
1. & x & & & = 10.35, \\
2. & -x & +y & & = 7.25, \\
3. & & y & & = 17.63, \\
4. & & y & -z & = 9.10, \\
5. & x & & -z & = 1.94.
\end{array}$$

Now to form the normal equation for x, the first equation is to be multiplied by 1, the second by -1, and the fifth by 1; and adding these,

$$3x - y - z = 5.04.$$

In like manner to find the normal equation for y, the second equation is multiplied by 1, the third by 1, and the fourth by 1, whence

$$-x + 3y - z = 33.98.$$

Lastly, to find the normal equation for z, the fourth equation is multiplied by -1 and the fifth by -1, and adding,

$$-x - y + 2z = -11.04.$$

These three normal equations contain three unknown quantities, and their solution gives

$$x = 10.372, \qquad y = 17.61, \qquad z = 8.47 \text{ feet},$$

which are the most probable values of the three elevations.

As a second example the three observation equations near the middle of the last Article are

$$\begin{aligned} x \qquad &= 62^\circ\ 20'\ 43'', \\ y &= 36\ \ 14\ \ 06, \\ x + y &= 98\ \ 34\ \ 52. \end{aligned}$$

Applying the rule, the two normal equations are

$$\begin{aligned} 2x + y &= 160\ \ 55\ \ 35, \\ x + 2y &= 134\ \ 48\ \ 58, \end{aligned}$$

and the solution of these gives

$$x = 62^\circ\ 20'\ 44'', \quad y = 36^\circ\ 14'\ 07'',$$

whence the third angle of the triangle is 180 degrees minus the sum of these, or $81^\circ\ 25'\ 09''$. By comparing these with the observed values it will be seen that each observation is corrected by the same amount; this is because the observations are of equal weight and each angle is similarly related to the other two.

Prob. 6. Form and solve the normal equations for the observation equations of Prob. 5.

7. Indirect Observations of Unequal Weight.

The two preceding Articles give the method of adjusting indirect observations of equal weight upon several independent quantities; now is to be investigated the case of indirect observations of different weights upon such quantities. Let $p_1, p_2, \ldots p_n$ be the weights of the n observations M_1, M_2, $\ldots M_n$, so that the observation equations are

$$a_1x + b_1y + c_1z + \ldots = M_1, \quad \text{with weight } p_1,$$
$$a_2x + b_2y + c_2z + \ldots = M_2, \quad \text{with weight } p_2,$$
$$\cdot \quad \cdot \quad \cdot \quad \cdot \quad \cdot \quad \cdot \quad \cdot \quad \cdot \quad \cdot$$
$$a_nx + b_ny + c_nz + \ldots = M_n, \quad \text{with weight } p_n.$$

Now if the first equation were written p_1 times, the second p_2 times, etc., all the equations would have the same weight and the rule of the last Article would apply. That is, if each of the above equations be multiplied by the coefficient of x in that equation, and also by its weight, the sum will be the condition for the most probable value of x; and in like manner is found the condition for the most probable value of each of the other unknowns. These conditions are the normal equations.

The following is hence the rule for the adjustment of observations of unequal weight upon several independent quantities:

> For each of the unknown quantities form a normal equation by multiplying each observation equation by the coefficient of that unknown quantity in that equation, and also by its weight, and adding the results. The solution of these normal equations will furnish the most probable values of the unknown quantities.

In applying this rule the same precautions are to be observed regarding signs of the coefficients as before stated.

An algebraic expression of the normal equations can be made by introducing the following abbreviations:

$$\begin{aligned}
[pa^2] &= p_1a_1^2 + p_2a_2^2 + \ldots + p_na_n^2, \\
[pab] &= p_1a_1b_1 + p_2a_2b_2 + \ldots + p_na_nb_n, \\
[paM] &= p_1a_1M_1 + p_2a_2M_2 + \ldots + p_na_nM_n, \\
&\cdot \quad \cdot \quad \cdot \quad \cdot \quad \cdot \quad \cdot \quad \cdot
\end{aligned}$$

and then the normal equations can be written

$$\left.\begin{aligned}
[pa^2]x + [pab]y + [pac]z + \text{etc.} &= [paM], \\
[pab]x + [pb^2]y + [pbc]z + \text{etc.} &= [pbM], \\
[pac]x + [pbc]y + [pc^2]z + \text{etc.} &= [pcM]. \\
\cdot \quad \cdot \quad \cdot \quad \cdot \quad \cdot \quad \cdot \quad \cdot &
\end{aligned}\right\} \qquad (7)$$

Here it will be seen that the coefficients of the unknown quantities in the first vertical column are the same as those in the first horizontal line, those in the second column the same as those in the second line, and so on. This is a characteristic of normal equations and serves as a check when they are deduced by direct application of the rule. If the observations are of equal weight, p is to be made unity throughout, and the method reduces to that of the last Article.

As a numerical illustration let five observations produce the five observation equations

$$\begin{array}{lrcll}
1. & +x & = & 0, & \text{with weight } 3, \\
2. & +y & = & 0, & \text{with weight } 19, \\
3. & +z & = & 0, & \text{with weight } 13, \\
4. & +x+y & = & +0.34, & \text{with weight } 17, \\
5. & +y+z & = & -0.18, & \text{with weight } 6.
\end{array}$$

From these the normal equations, formed either by the rule or by help of the algorithm, are

$$\begin{aligned}
20x + 17y \qquad &= +5.78, \\
17x + 42y + 6z &= +4.70, \\
6y + 19z &= -1.08,
\end{aligned}$$

whose solution furnishes the results

$$x = +0.285, \quad y = +0.005, \quad z = -0.059,$$

which are the most probable values of the required quantities.

Prob. 7. In a plane triangle six observations give $A = 42^\circ\ 17'\ 35''$, three observations give $B = 56^\circ\ 40'\ 09''$, and two observations give $C = 81^\circ\ 02'\ 10''$. Compute the adjusted values of the angles.

8. Solution of Normal Equations.

The normal equations which arise in the adjustment of observations may be solved by any algebraic process. It is desirable, however, to use methods which will furnish the value of each unknown quantity independently of the others, as the liability to error is thus lessened. When there are but two normal equations let them be expressed in the form

$$A_1x + B_1y = D_1,$$
$$A_2x + B_2y = D_2,$$

then the solution by any method gives the formulas

$$x = \frac{B_1D_2 - B_2D_1}{B_1A_2 - B_2A_1}, \quad y = \frac{A_1D_2 - A_2D_1}{A_1B_2 - A_2B_1},$$

which can easily be kept in mind by noting the order of the letters and subscripts. It may be observed also that the two denominators are equal numerically but of contrary sign.

For three normal equations let them be written in the form

$$A_1x + B_1y + C_1z = D_1,$$
$$A_2x + B_2y + C_2z = D_2,$$
$$A_3x + B_3y + C_3z = D_3,$$

and the solution leads to the formulas

$$\left.\begin{aligned} x &= \frac{(B_2C_3-B_3C_2)D_1+(B_3C_1-B_1C_3)D_2+(B_1C_2-B_2C_1)D_3}{(B_2C_3-B_3C_2)A_1+(B_3C_1-B_1C_3)A_2+(B_1C_2-B_2C_1)A_3}, \\ y &= \frac{(A_3C_2-A_2C_3)D_1+(A_1C_3-A_3C_1)D_2+(A_2C_1-A_1C_2)D_3}{(A_3C_2-A_2C_3)B_1+(A_1C_3-A_3C_1)B_2+(A_2C_1-A_1C_2)B_3}, \\ z &= \frac{(A_2B_3-A_3B_2)D_1+(A_3B_1-A_1B_3)D_2+(A_1B_2-A_2B_1)D_3}{(A_2B_3-A_3B_2)C_1+(A_3B_1-A_1B_3)C_2+(A_1B_2-A_2B_1)C_3}, \end{aligned}\right\} \quad (8)$$

in which the three denominators have the same value. After a little practice it will be easy to use these formulas with great rapidity in the solution of normal equations.

When the number of normal equations is greater than three, general formulas for solution are too lengthy to be written, and the systematic method of substitution devised by Gauss is generally employed. This is explained and exemplified in text-books on the Method of Least Squares, but lack of space forbids its presentation here.

Prob. 8. Solve the normal equations

$$3x - y + 2z = 5, \quad -x + 4y + z = 6, \quad 2x + y + 5z = 3,$$

and check the solution by showing that the values of the roots satisfy the equations.

9. The Probable Error.

The Method of Least Squares comprises two tolerably distinct divisions. The first is the adjustment of observations, or the determination of the most probable values of observed quantities. The second is the investigation of the precision of observations and of the adjusted results. The first is done by the application of the principle of least squares given in Art. 3; the second is done by the determination of the probable error, the rules for which will now be presented.

The following may be stated as a definition of the term "probable error":

> In any large series of errors the probable error is an error of such a value that the number of errors less than it is the same as the number greater than it.

The probable error is hence an error which is as likely as not to be exceeded. In the figure if the ordinate MN be drawn so as to divide the area on each side of OY into two equal parts, then OM is the probable error. Here the total area between the curve and the X-axis is unity (certainty), and the area $MNYNM$ is 0.5; thus the probability that an error is greater than OM is 0.5, and that it is less than OM is also 0.5.

To render more definite the conception of probable error let two sets of observations made upon the length of a line

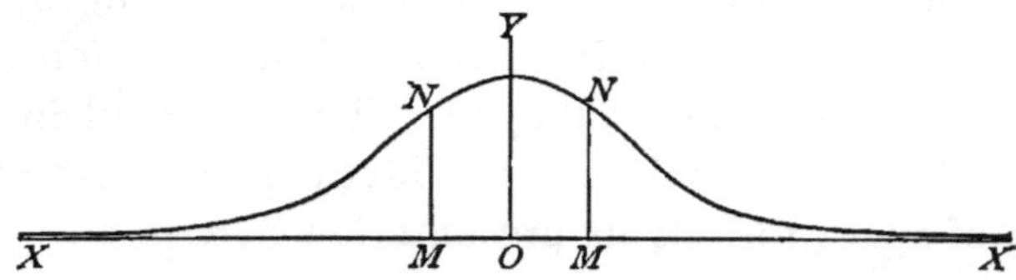

be considered. The first set, made with a chain, gives 634.7 feet with a probable error of 0.3 feet. The second set, made with a tape, gives 634.64 with a probable error of 0.06 feet; thus,

$$l_1 = 634.7 \pm 0.3 \quad \text{and} \quad l_2 = 634.64 \pm 0.06;$$

and it is an even chance that 634.7 is within 0.3 of the truth, and also an even chance that 634.64 is within 0.06 of the truth. The probable error thus gives an absolute idea of the accuracy of the results; it also serves as a means of comparing the precision of different observations, for in the above case the precision of the second result is to be taken as much greater than that of the first.

It is a principle of the Method of Least Squares that weights of observations are reciprocally proportional to the squares of their probable errors. Thus, for the above numerical example,

$$p_1/p_2 = \frac{1}{0.09}\Big/\frac{1}{0.0036} = 1/25.$$

Hence the second observation has a value about 25 times that of the first when it is to be used in combination with other measurements. Weights and probable errors are constantly used in the discussion of observations. Weights are usually determined from the number of measurements or from knowledge of the manner in which they are made, but probable errors are computed from the observations themselves.

For the case of direct observations on the same quantity, all being of equal precision, the arithmetic mean is the most probable value (Art. 3). Subtracting each observation from the mean gives the residuals $v_1, v_2, \ldots v_n$, and the sum of the squares of these is represented by Σv^2. Then

$$r_1 = 0.6745\sqrt{\frac{\Sigma v^2}{n-1}} \tag{9}$$

is the probable error for a single observation, and, since n is the weight of the arithmetic mean,

$$r = \frac{r_1}{\sqrt{n}} \tag{9}'$$

is the probable error of the arithmetic mean. For example, let six observations of an angle be taken with equal care and let these be arranged as below in the column headed M. The sum of these values divided by 6 gives 48° 06′ 14″.7 as

M	v	v^2	
48° 06′ 12″.5	+2″.2	4.84	
15 .0	−0 .3	0.09	$r_1 = 2''.58 \quad r = 1''.05$
20 .3	−5 .6	31.36	
08 .9	+5 .8	33.64	$z = 48°\ 06'\ 14''.7 \pm 1''.1$
15 .1	−0 .4	0.16	
16 .4	−1 .7	2.89	
$z = 48°$ 06′ 14″.7	0″.0	72.98 = Σv^2	

the most probable value of the angle, the second column gives the residuals, and the third their squares. Then by the use of the formula the probable error of a single observation is found to be 2″.58 and that of the arithmetic mean to be 1″.05. Thus if another observation were to be taken it is as likely as not that it will deviate 2″.58 from the truth.

For the case of n observations of different weights on one quantity the weighted mean is the most probable value (Art. 9). Subtracting each observation from this gives the residuals, and the square of each of these is to be multiplied by its weight to give the sum of the weighted squares, which may be represented by Σpv^2. Then

$$r_1 = 0.6745 \sqrt{\frac{\Sigma pv^2}{n-1}} \qquad (9)''$$

is the probable error of an observation of the weight unity, and if Σp represent the sum of the weights,

$$r = \frac{r_1}{\sqrt{\Sigma p}}$$

is the probable error of the weighted mean. As an example, let the observations in the first column of the following table be the results of the repetition of an angle at different times, 18″.26 arising from five repetitions, 16″.30 from four, and so on, the weights of the observations being taken the same as the number of repetitions. Then the general mean z has

M	p	v	v^2	pv^2
32° 07′ 18″.26	5	− 0″.10	0.010	0.05
16 .30	4	+ 1 .86	3.460	13.84
21 .06	1	− 2 .90	8.410	8.41
17 .95	4	+ 0 .21	0.044	0.18
16 .20	3	+ 1 .96	3.842	11.53
20 .85	4	− 2 .69	7.236	28.94
$z =$ 32° 07′ 18″.16	$21 = \Sigma p$			$\Sigma pv^2 = 62.95$

the weight 21, the sum of the several weights or the number of single measures. Subtracting each M from z gives the residuals in the column v; next from a table of squares the numbers in the column v^2 are found, and multiplying each of these by the corresponding weight gives the quantities pv^2 whose sum is 62.95. Then, since n is 6, the probable error of an observation whose weight is unity is found from the formula to be $r_1 = 2''.39$ and that of the weighted mean to be $r = 0''.52$. Hence the final value of the angle may be written $z = 32° \; 07' \; 18''.16 \pm 0''.52$, which indicates a high degree of precision.

Prob. 9. Four measurements of a base line give the results 922.220 feet, 922.197 feet, 922.221 feet, and 922.217 feet. Compute the probable error of the most probable value.

10. Probable Errors for Indirect Observations.

It is sometimes required to find the probable errors of the observed quantities $M_1, M_2, \ldots M_n$, and the probable errors of the quantities x, y, z, etc., whose values have been obtained by the methods of Arts. 7 and 8. These may be found by first deducing the probable error of an observation of the weight unity and then dividing this by the weights $p_1, p_2, \ldots p_n$ and p_x, p_y, p_z, etc. If n is the number of observations, q the number of unknown quantities, and Σpv^2 the sum of the weighted squares of the residuals, then, as shown in treatises on the Method of Least Squares,

$$r_1 = 0.6745\sqrt{\frac{\Sigma pv^2}{n-q}} \qquad (10)$$

is the formula for the probable error of an observation of the weight unity, and

$$r_1 = \frac{r_1}{\sqrt{p_1}}, \qquad r_x = \frac{r_1}{\sqrt{p_x}},$$

are the probable errors of M_1 and of x respectively.

The weights $p_1, p_2, \ldots p_n$ are known, but the weights p_x, p_y, etc., are to be derived by preserving the absolute terms of the normal equations in literal form during the solution. Then the weight of any unknown quantity is the reciprocal of the coefficient of the absolute term which belongs to the normal equation for that unknown quantity. For instance, take the normal equations

$$\begin{aligned} 3x - y - z &= D_1, \\ -x + 3y - z &= D_2, \\ -x - y + 2z &= D_3. \end{aligned}$$

The solution of these by any method gives

$$x = \tfrac{5}{8}D_1 + \tfrac{3}{8}D_2 + \tfrac{1}{2}D_3,$$
$$y = \tfrac{3}{8}D_1 + \tfrac{5}{8}D_2 + \tfrac{1}{2}D_3,$$
$$z = \tfrac{1}{2}D_1 + \tfrac{1}{2}D_2 + D_3.$$

Hence the weight of x is $\frac{8}{5}$, that of y is $\frac{8}{5}$, and that of z is 1. If it be only desired to find the weight of x, the terms D_2 and D_3 need not be retained in the computation; if only to find the weight of z, the terms D_1 and D_2 can be omitted.

As a numerical example the observation equations given at the beginning of Art. 5 may again be considered. These may be written, if x, y, and z denote the most probable elevations,

$$\begin{aligned} x - 10.35 &= v_1, \\ y - x - 7.25 &= v_2, \\ y - 17.63 &= v_3, \\ y - z - 9.10 &= v_4, \\ x - z - 1.94 &= v_5, \end{aligned}$$

in which v_1, v_2, etc., are the residual errors. Now in Art. 6 the most probable values were derived,

$$x = 10.37, \quad y = 17.61, \quad \text{and} \quad z = 8.47 \text{ feet},$$

and substituting these, the residuals are found to be

$$v_1 = +0.02, \; v_2 = -0.01, \; v_3 = -0.02, \; v_4 = +0.04, \; v_5 = -0.04.$$

Now, as the weights are equal, Σpv^2 becomes Σv^2, and its value is $\Sigma v^2 = 0.0041$. Then, since n is 5 and q is 3,

$$r_1 = 0.6745\sqrt{\frac{0.0041}{5-3}} = 0.031 \text{ feet},$$

which is the probable error of a single observation. By the method above explained it will be found that the weight of x is 1.8, whence its probable error is

$$r_x = \frac{0.031}{\sqrt{1.8}} = 0.023 \text{ feet},$$

and in a similar manner the probable errors of y and z are 0.023 feet and 0.031 feet. The final adjusted values may then be written

$$x = 10.37 \pm 0.02, \quad y = 17.61 \pm 0.02, \quad z = 8.47 \pm 0.03.$$

Prob. 10. Four measurements give the observation equations

$$\begin{aligned} +x \qquad\qquad &= 12.27, \quad \text{with weight } 2, \\ -x + y \qquad &= 1.04, \quad \text{with weight } 2, \\ -y + z &= 3.30, \quad \text{with weight } 1, \\ z &= 16.67, \quad \text{with weight } 1. \end{aligned}$$

Find the most probable values of x, y, and z, their weights and their probable errors.

11. Probable Errors of Computed Values.

The determination of the precision of quantities which are computed from observed quantities is now to be discussed. For instance, the area of a field is computed from its sides and angles; when the most probable values of these have been found by measurement, the most probable value of the area is computed by the rules of geometry, and the precision of that area will depend upon the precision of the observed quantities.

Let z_1 and z_2 be two adjusted values whose probable errors are r_1 and r_2; it is required to find the probable error r of the sum $z = z_1 + z_2$. If v_1', v_1'', etc., be residual errors for z_1 and v_2', v_2'', etc., be residual errors for z_2, then the corresponding errors for z are $v' = v_1' + v_2'$, $v'' = v_1'' + v_2''$, etc. Squaring each of these and adding the results gives

$$\Sigma v^2 = \Sigma v_1^2 + 2\Sigma v_1 v_2 + \Sigma v_2^2,$$

and for a large number of errors $\Sigma v_1 v_2$ is zero, since each product $v_1 v_2$ is as likely to be positive as negative. Now Σv^2, Σv_1^2, and Σv_2^2 are proportional to r^2, r_1^2, and r_2^2 as seen by (9), and accordingly

$$r = \sqrt{r_1^2 + r_2^2}$$

gives the probable error of the sum $z_1 + z_2$. In like manner it may be shown that the probable error of the difference $z_1 - z_2$ is also given by $\sqrt{r_1^2 + r_2^2}$. Further, if $z = z_1 \pm z_2 \pm \ldots \pm z_n$, then

$$r^2 = r_1^2 + r_2^2 + \ldots + r_n^2 \qquad (11)$$

determines the probable error of z. For example, if a base line be measured in three parts giving 250.33 ± 0.05, 461.29 ± 0.07, and 732.40 ± 0.10 feet, then $r = 0.13$ feet, and the total length may be written 1444.02 ± 0.13 feet.

If x be an observed quantity whose probable error is r, then the probable error of ax is ar. Thus, if the diameter of a circle be observed to be 42 feet 2 inches ± 0.5 inches, the circumference is 132.47 ± 0.13 feet.

If X be any function of x, then the error dx in x produces the error dX in x, and the error r in x produces the error $r\frac{dX}{dx}$ in X. For example, let x be the observed diameter of a circle and r its probable error; then $X = \frac{1}{4}\pi x^2$ is its area, and $dX = \frac{1}{2}\pi x \cdot dx$, whence the probable error of X is $r \cdot \frac{1}{2}\pi x$. Thus, if x is 42 feet 2 inches ± 0.5 inches, the area is 1396.46 ± 2.76 square feet.

Lastly, let X be any function of the independently observed quantities x, y, z, etc., and let it be required to find the probable error of X from the probable errors r_1, r_2, r_3, etc., of the observed quantities. If the measurements are made with precision, so that the probable errors are small, it can be shown that

$$r^2 = \left(r_1\frac{dX}{dx}\right)^2 + \left(r_2\frac{dX}{dy}\right)^2 + \left(r_3\frac{dX}{dz}\right)^2 + \ldots \qquad (11)'$$

determines the probable error of X. For example, let x and y be the sides of a rectangular field and $X = xy$ its area. Then the probable error r_1 in x gives the probable error $r_1 y$ in X, and the probable error r_2 in y gives the probable error

$r_2 x$ in X, so that $(r_1 y)^2 + (r_2 x)^2$ is the square of the resulting probable error of X. Thus, if $x = 50.00 \pm 0.01$ feet and $y = 200 \pm 0.02$ feet, the area is $10\,000 \pm 2.24$ square feet.

Formula (11)′ will be frequently used in the following pages, it being a general rule that includes all cases. As another illustration let A and B be two points whose horizontal distance apart is l, and let θ be the vertical angle of elevation of B above A; let r_1 be the probable error of l, and r_2 the probable error of θ. The height of B above A is given by $X = h \tan\theta$, and, by the application of the formula, regarding h as x and θ as y, there results

$$r^2 = (r_1 \tan\theta)^2 + (r_2 l/\cos^2\theta)^2.$$

If $l = 1035.2 \pm 1.3$ feet, and $\theta = 3° \; 10' \pm 02'$, then $r_1 = 1.3$ feet, but, to make the computation, r_2 must be expressed in the same unit as $\cos^2\theta$, that is, in radians; since 3438 minutes make one radian, the numerical value of r_2 is 2/3438. Then are found $r_1 \tan\theta = 0.072$ feet, $(r_2 l/\cos^2\theta) = 0.604$ feet, whence $r = 0.608$ feet. The value of X being $1035.2 \tan 3° \; 10' = 57.27$ feet, this may be written 57.27 ± 0.61 feet, and thus it is as likely as not that the error in the computed height is less than 0.61 feet. Here it is seen that the probable error in the small vertical angle produces the greater part of the probable error in the computed result.

Prob. 11. In a plane triangle ABC let $A = 90°$, $C = 16° \; 04' \; 45'' \pm 30''$, and $a = 6256.8 \pm 0.7$. Compute the length of the side c and its probable error.

12. Critical Remarks.

The most important processes for the adjustment and comparison of observations have now been presented, but the brief space at command has forbidden extended theoretic discussions like those found in treatises on the Method of Least Squares. The student has been obliged to take for-granted the law of probability of error and the formulas for probable errors, but otherwise the subject has been developed in logical manner. Legendre, in announcing the principle of least squares in 1805, gave no proof of its correctness or validity; he notes, however, that this principle balances the errors, so that the effect of the extreme ones is neutralized.

In mechanics the center of gravity is a point about which all the particles of the body balance; so the arithmetic mean gives a value about which all the errors balance, the sum of their residuals being zero. The moment of inertia of a body is a minimum for an axis passing through the center of gravity; so the sum of the squares of the residual errors is to be made a minimum in order to find the most probable values of an observed quantity. The radius of gyration with respect to an axis through the center of gravity bears also an analogy to the probable error. Thus the Method of Least Squares may be justified by the mechanical principles of equilibrium.

Numerous applications of the adjustment of observations will be given in the following Chapters, and a simplification will be introduced whereby the formation of normal equations from observation equations may be rendered numerically easier. A treatment of conditioned observations by the use of "correlate equations" will also be presented, whereby the work of computation may often be materially shortened. As measurements become more and more precise the necessity for rational processes of adjustment and comparison becomes greater and greater. In physics, astronomy, geodesy, and wherever precise observations are taken, the Method of Least Squares is now universally used, and there is little doubt but that in future years all books on surveying will treat more or less of its principles and processes.

A list of writings on errors of observations and on the Method of Least Squares from 1722 to 1876 will be found in Transactions of the Connecticut Academy, 1877, vol. IV, pp.

151–222. Many of these, together with others from 1877 to 1888, are given in Gore's Bibliography of Geodesy, published in Report of the U. S. Coast and Geodetic Survey for 1887, pp. 313–512.

Prob. 12. A base line was measured in three parts, the values found for these being 126.74, 219.18, and 270.40 meters. The total length was then measured and found to be 616.39 meters. Find the adjusted length of the base, the weights of the four observations being 17, 9, 8, and 3.

CHAPTER II.

PRECISE PLANE TRIANGULATION.

13. COORDINATES AND AZIMUTHS.

Plane surveying is that which covers an area so small that it is unnecessary to take into account the curvature of the earth's surface. Surveys of cities, townships, harbors, and mines are usually of this character. The field operations of the plane triangulations of such surveys do not differ in principle from those of geodetic triangulation, the latter being merely more precise.

In geography the position of a point on the earth's surface is located by its angular distance north or south of the equator and by its angular distance east or west of the meridian of Greenwich, these coordinates being called latitude and longitude. In plane surveying two straight lines are imagined to be drawn at right angles to each other, one coinciding with the meridian, and these constitute a system of coordinate axes to which points are referred by rectangular coordinates. The linear distance of a point east or west of the meridian is called its longitude, and the linear distance north or south from the other axis is called its latitude. The coordinate axes are rarely laid out on the ground, but upon the maps they are drawn, as also lines parallel to them at regular distances apart, thus forming a system of squares by which points are readily located.

The azimuth of a line AB is the angle that it makes with a meridian drawn through the end A. Azimuths are usually measured around the circle from 0° to 360°; thus if the azimuth of AB is 40° the azimuth of BA is 220° in plane surveying.

There are in use several systems of reckoning coordinates and azimuths. The one most commonly used in plane surveying has the latitudes positive when measured north and negative when measured south, while the longitudes are positive toward the east and negative toward the west. In this system azimuths are reckoned around from the north through the east, the azimuth of north being 0°, that of east 90°, that of south 180°, and that of west 270°. This system is used in the Handbook for Surveyors.

In geodetic surveying in America latitudes and longitudes are reckoned as in geography, north latitude being positive and south latitude negative, while west longitude is positive

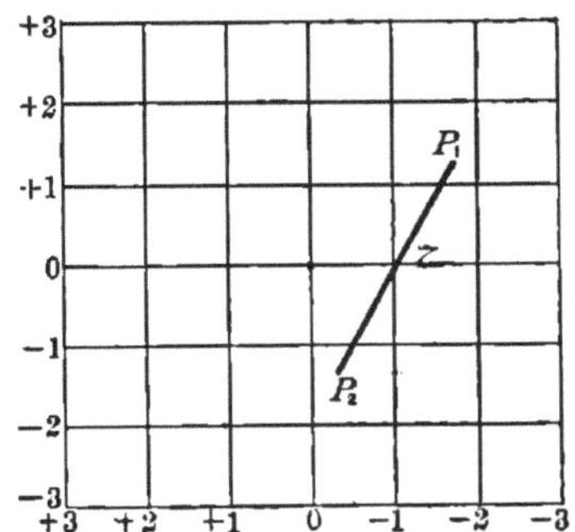

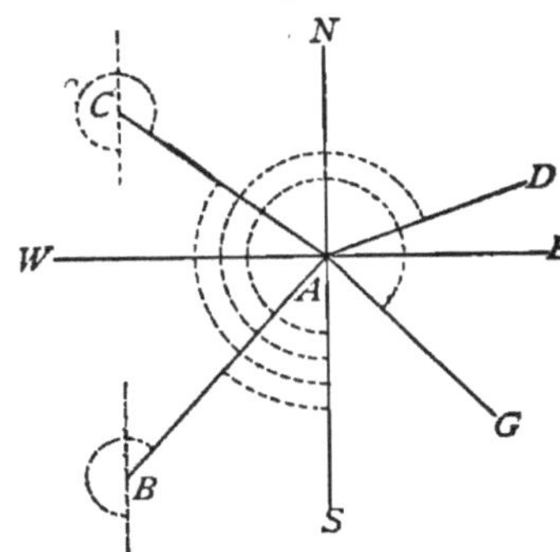

and east longitude is negative. Here the azimuths are reckoned from the south around through the west, the azimuth of south being 0°, that of west being 90°, that of north 180°, and that of east 270°. This system is also employed for the linear coordinates of plane surveys based on geodetic triangulations, and it will be used throughout this volume. For a city survey the origin may be taken through the tower of the city hall, or, if it is desired to avoid negative latitudes and longitudes, it may be taken near the southeast corner of the city. The size of the squares drawn upon the map will depend upon its scale; the side of a square is usually taken as 1 000 feet or meters, or 10 000 feet or meters.

In thinking of the azimuth of a line the student should imagine himself to be standing at the end which is first mentioned in its name and to be looking toward the other end; then he should imagine a meridian drawn through that end toward the south, and the angular deviation of these lines, measured as above described, is the azimuth. Thus, let the azimuth of *AC* in the above figure be 115°, then the azimuth of *CA*, determined by drawing a meridian through *C*, is 115° + 180° or 295°. For all cases in plane surveying the back azimuth of a line is 180° greater than the front azimuth, because the meridians are parallel.

When several lines radiate from a station and their azimuths are known the angle between any two lines is found by taking the difference of their azimuths. Thus, if the azimuth of *AB* is 35° 17′ 04″ and that of *AC* is 120° 46′ 19″ the angle *BAC* is 85° 29′ 15″. Again, for the lines *AB* and *AG* let the azimuth of the first be as before and that of *AG* be 320° 10′ 02″, then the angle *GAB* is 75° 06′ 52″; here 360° is to be added to the azimuth of *AB* before subtracting from it the azimuth of *AG*.

Let P_1 and P_2 be two points in such a plane coordinate system, L_1 and L_2 their latitudes, M_1 and M_2 their longitudes, l the length of the line joining the points, and Z the azimuth of P_1P_2. Then, if L_1, M_1, l, and Z are given, the coordinates of P_2 are

$$L_2 = L_1 - l\cos Z, \quad M_2 = M_1 + l\sin Z, \qquad (13)$$

which are always correct if $\cos Z$ and $\sin Z$ are used with their proper signs according to the value of Z. For example, let $L_1 = +20\,148.3$ feet, $M_1 = +45\,933.7$ feet, $l = 7\,789.5$ feet, $Z = 205°\ 36'\ 07''$; here both $\cos Z$ and $\sin Z$ are negative, and the computation gives $L_2 = +27\,173.0$ feet and $M_2 = +42\,567.7$ feet for the coordinates of the second point.

A triangulation is a cheap and accurate method for determining the coordinates of stations. The stations are first located on the ground so as to give good-shaped triangles, two of them being so placed as to form a base line whose length can be measured with precision. The angles of all the triangles are then observed, and from these and the length of the base the lengths of all the sides are computed. The azimuth of one of the sides is determined by astronomical observations, and from this and the angles the azimuth of each side is known. The coordinates of one of the stations being assumed, the coordinates of all other stations are computed. Lastly, the lengths and azimuths of the sides and the coordinates of the stations are recorded as the basis for topographic surveys, and the coordinate system being plotted the stations are laid down in their correct relative positions.

In this Chapter those operations of plane triangulation will be discussed which depend upon the measurement of horizontal angles. Strictly speaking triangulation includes base-line and azimuth observations, as these must be made before the angle work can be fully computed. It will be convenient, however, to first discuss the angular measurements and their adjustments, leaving the base lines to be treated in Chapter III and the azimuth observations in Chapter V.

Prob. 13. Let the latitude and longitude of Q be $+ 6\,131.31$ meters and $+ 36\,414.60$ meters, the length QN be $12\,454.02$ meters, and the azimuth of QN be $300° 06' 31''$. Draw the figure and compute the latitude and longitude of N.

14. Measurement of Angles.

Horizontal angles are measured either with a direction instrument or with a repeating instrument. A direction instrument has no verniers, but the readings are made by several micrometer microscopes placed around the graduated circle. Any engineer's transit may be used as a repeating instrument, and the following notes will treat of work done with these. A good transit, having two verniers reading to half-minutes, can easily measure horizontal angles with a probable error of one second if proper precautions be taken to eliminate systematic and accidental errors.

Errors due to setting the transit or the signals in the wrong position cannot be eliminated, and hence great care must be taken that they are centered directly over the stations. If the graduated limb be not horizontal the measured angles will be always too large and hence the levels on the limb must be kept in true adjustment. All the adjustments of the transit, in fact, must be carefully made and preserved in order to secure precise work.

Errors due to inaccurate setting of the verniers, as also those due to eccentricity between the center of the alidade and the center of the limb, may be eliminated by reading both verniers and taking the mean. Errors due to collimation and to a difference in height of the telescope standards may be eliminated by taking a number of measures with the telescope in its normal or direct position and an equal number with it in the reverse position. Errors due to inaccurate graduation may be eliminated by taking readings on different parts of the circle. Errors due to pointing and to clamping may be largely eliminated by taking one half of the measures from left to right and the other half from right to left. Lastly, in order to eliminate errors due to atmospheric influences it is well to take different series of measurements on different days.

The following form of field notes shows four sets of measurements of an angle HOK, each set having three repetitions. The first and fourth sets are taken with the telescope in the direct position, the second and third with it reversed. The first and second sets are taken by pointing first at H and next at K, the third and fourth are taken by pointing first at K and next at H. At each reading both verniers are noted.

The vernier is never set at zero, but the reading before beginning a set is usually made to differ by about 90° from that of the preceding set so as to distribute the readings uniformly over the circle. In the first and second sets the mean final reading minus the mean initial reading is divided by 3 to give the angle; in the third and fourth sets the mean

FIELD NOTES OF HORIZONTAL ANGLE *HOK*.

Station observed.	No. of Reps.	Tel. D. or R.	Reading.					Angle.	Remarks.
			°	′	A ″	B ″	Mean ″	° ′ ″	
H	3	*D*	10	02	00	30	15.0	62 25 10.0	Angle at Station O.
K			197	17	30	60	45.0		Sept. 31, 1895, P.M.
H	3	*R*	100	11	30	30	30.0	62 25 07.5	Brandis Transit, No. 716
K			287	26	60	45	52.5		John Doe, observer. R. Roe, recorder.
K	3	*R*	190	01	30	45	37.5	62 25 23.3	Clear, air hazy.
H			2	45	15	40	27.5		
K	3	*D*	280	55	45	60	52.5	62 25 30.0	Mean of 4 sets
H			93	39	00	45	22.5		$HOK = 62°25' 17''.7$

initial reading minus the mean final reading is divided by 3. The mean of the four values of the angle is 62° 25′ 17″.7, which is its most probable value as determined by these observations.

In repeating angles the following points should be noted. The transit should never be turned upon its vertical axis by taking hold of the telescope or of any part of the alidade. The limb should never be clamped when the verniers are read. The observer should not walk around the transit to read the verniers, but standing where the light is favorable he should revolve the limb so as to bring vernier *A* before him and then vernier *B*. The observer should not allow his knowledge of the reading of vernier *A* to influence him in taking that of *B*. Care must be taken to turn the clamps slowly and not too tightly. If these precautions be taken, and if the observer becomes skilled in manipulation and close reading of the verniers, it will be possible to obtain the value of an angle to a high degree of precision with a transit reading only to minutes.

Four sets taken in the manner just described constitute a series. The number of series required will depend upon the precision demanded in the work. If it be required to render the probable error of the final result about one second, it will generally be necessary to take about 6 or 8 series. By taking these in one day a smaller probable error may be found than if they were taken on two or more different days, but the final result will really be more precise in the latter case because it eliminates numerous errors due to atmospheric influences.

When three lines meet at a station there are but two independent angles to be observed, when four lines meet there are but three independent angles, and in general n lines give $n - 1$ independent angles. It is, however, generally best to

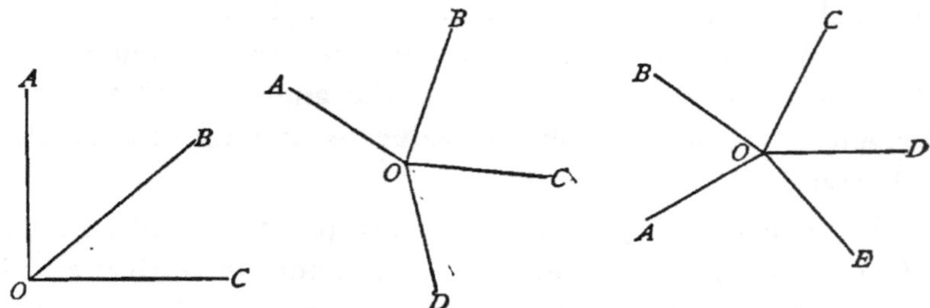

measure all the angles resulting from the combination of the lines two by two. Thus for three lines OA, OB, and OC, the angles AOB, AOC, and BOC should be observed; for four lines six angles should be measured, namely, AOB, AOC, AOD, BOC, BOD, and COD; in general for n lines the measurements should be distributed over $\frac{1}{2}n(n - 1)$ angles. If about 5 series are thought necessary for one independent angle, then for four independent angles 20 series are required, but if these are distributed over the ten combined angles then only 2 series need be taken upon each. The adjustment of the observed values by the Method of Least Squares gives finally the most probable values of all the angles at the station.

Another method, which is used by many observers, is to measure each of the n angles included between the n lines; thus for the four lines in the middle diagram the angles AOB, BOC, COD, and DOA would be measured. The theoretic sum of these being 360 degrees, the observations are then to be adjusted to agree with this condition and at the same time render the sum of the squares of the residuals a minimum (Art. 21).

In writing the letters designating an angle it is desirable to do so in the order of azimuths, that is, standing at the vertex of the angle the letter on the left-hand line should be mentioned first. Thus AOB means the angle from the line OA around to OB, but BOA means the angle from OB around to OA and it is, of course, 360° minus AOB. By this method an angle is estimated in the same direction around the circle as is azimuth, and thus uniformity is secured and ambiguity avoided.

Prob. 14. Show that the probable error of the value of HOK found from the above field notes is $3''.6$.

15. Probable Errors and Weights of Angles.

In the field note-book the observations are recorded in the order in which they are made, but it is desirable before the occupation of a station is concluded that the results for each angle should be arranged in an abstract and the probable error be computed. Thus the observer gains a clear idea of the precision of each angle and is able to decide whether additional measures are necessary. The weights of the final means are, however, usually assigned from the number of repetitions rather than by the probable errors.

The following is an abstract of the observations of an angle *PNE* measured on the precise triangulation around Lehigh University in 1898, each result being the mean of four sets taken in the manner shown in the field notes of Art. 15, and

ABSTRACT OF HORIZONTAL ANGLES.

Date. 1898.	No. of Reps.	Angle *PNE*.	v	v^2	Remarks.
Oct. 3	12	12° 15′ 19″.3	− 2.9	8.4	Buff and Berger Transit.
Oct. 4	12	13 .8	+ 2.6	6.8	
Oct. 10	12	16 .3	+ 0.1	0.0	Each series taken by a different observer.
Oct. 11	12	21 .8	− 5.4	29.2	
Oct. 17	12	09 .5	+ 6.9	47.6	
Oct. 18	12	17 .6	− 1.2	1.4	$r_1 = 2''.90$
	$p = 72$	12° 15′ 16″.4	$\Sigma v^2 = 93.4$		$r = 1''.18$

each being taken by a different observer. Here, proceeding as in Art. 9, the arithmetic mean of the six observations gives 12° 15′ 16″.4 as the most probable value of the angle *PNE*. The column headed v gives the residuals found by subtracting each observation from that mean, and then the sum of their squares is found to be 93.4. From (9) the probable error of a single result is computed to be 2″.90 and from (9)′ the probable error of the mean is 1″.18, which shows a good degree of precision considering that the observers were not experienced and that the transit reads only to minutes.

A young observer is usually tempted, after having computed the mean and found the probable errors, to reject some of the observations which have the largest residuals, in order thereby to apparently increase the precision of the results. This temptation must be resisted, as an unwarranted rejection is equivalent to a dishonest alteration of field notes.

There are, however, two cases where an observed value may properly be rejected, namely, if it is evidently a mistake, as when the degrees and minutes of the angle are wrong, and if a remark in the note-book shows it was taken under unfavorable conditions. Some observers allow themselves the liberty of rejecting an observation when its residual is greater than five or six times the computed probable error of a single observation. There are some reasons in favor of this practice, but more observations than one should never be rejected in this way.

Although the weights of observations are inversely proportional to the squares of their probable errors, it is found that it is better and more convenient to give weights to angles from the number of repetitions or series which produce them rather than from their computed probable errors. If the number of observations were large in each case the two methods might closely agree, but in ordinary practice they do not. An observer of much skill and experience may be allowed to assign weights to his angles with regard both to the number of repetitions and to the probable errors, but in general it has been found best to make the weights closely proportional to the number of repetitions provided the measurements are taken under the same conditions, that is, by observers and instruments of equal precision.

To illustrate let $PNE = 12° \; 15' \; 16''.4 \pm 1''.2$ as found by 6 series, $PNF = 35° \; 07' \; 42''.5 \pm 4''.8$ as found from 4 series, and $ENF = 22° \; 52' \; 24''.0 \pm 2''.4$ as found from 6 series. Here the probable errors indicate that the precision of PNE is much greater than that of ENF, but in making the adjustment it is best to take their weights as equal since each has been found from the same number of measures. Thus the weights of the three observations should be taken as 6, 4, and 6, or as 3, 2, and 3 in making the adjustment.

Prob. 15. Show that the adjusted values of the above observations are $PNE = 12° \; 15' \; 17''.0$, $PNF = 35° \; 07' \; 41''.6$, and $ENF = 22° \; 52' \; 24''.6$.

16. The Station Adjustment.

When several angles have been measured at a station they are to be adjusted by the methods of Arts. 6 and 7. It is here only necessary to give additional examples and to explain an abridgment whereby the numerical work is simplified.

As an example involving equal weights let the data be the same as on page 18, the five observation equations being

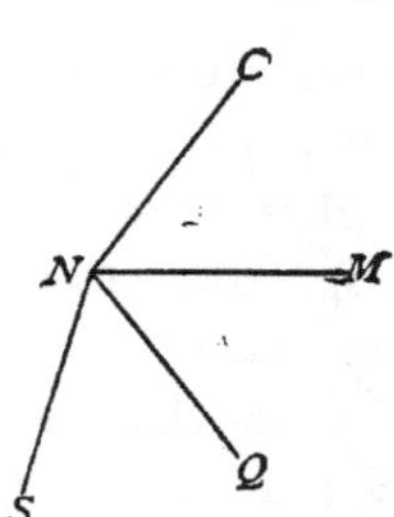

$$\begin{aligned}
+x \qquad\quad &= 55^\circ\, 57'\, 58''.68,\\
+y \quad\; &= 48\;\; 49\;\; 13\;.64,\\
+x+y \quad\; &= 104\;\; 47\;\; 12\;.66,\\
+z &= 54\;\; 38\;\; 15\;.53,\\
+y+z &= 103\;\; 27\;\; 28\;.99,
\end{aligned}$$

in which x, y, and z represent the angles CNM, MNQ, and QNS. Now let x_1, y_1, and z_1 be the most probable corrections to the measured values of x, y, and z, so that

$$\begin{aligned}
x &= 55^\circ\, 57'\, 58''.68 + x_1,\\
y &= 48\;\; 49\;\; 13\;.64 + y_1,\\
z &= 54\;\; 38\;\; 15\;.53 + z_1,
\end{aligned}$$

represent the most probable values of the quantities x, y, and z. Then substituting these in the observation equations the latter become

$$\begin{aligned}
+x_1 \qquad\quad &= \quad 0''.00,\\
+y_1 \quad\; &= \quad 0\;.00,\\
+x_1+y_1 \quad\; &= +0\;.34,\\
+z_1 &= \quad 0\;.00,\\
+y_1+z_1 &= -0\;.18.
\end{aligned}$$

Next, by the rule of Art. 6, the normal equations are

$$\begin{aligned}
2x_1 + \;y_1 \qquad\quad &= +0.34,\\
x_1 + 3y_1 + \;z_1 &= +0.16,\\
y_1 + 2z_1 &= -0.18,
\end{aligned}$$

the solution of which gives the corrections

$$x_1 = +0''.15, \quad y_1 = +0.04, \quad z_1 = -0.11,$$

and hence the most probable values of x, y, and z are

$$\begin{aligned} x &= 55^\circ\ 57'\ 58''.83 = CNM, \\ y &= 48\ \ 49\ \ 13\ .68 = MNQ, \\ z &= 54\ \ 38\ \ 15\ .42 = QNS, \end{aligned}$$

and from these by addition the most probable values of the other observed angles are

$$\begin{aligned} x + y &= 104\ \ 47\ \ 12\ .51 = CNQ, \\ y + z &= 103\ \ 27\ \ 29\ .10 = MNS. \end{aligned}$$

The residuals for the five observation equations, found by substituting the most probable values, are $+0''.15$, $+0''.04$, -0.15, -0.11, $+0.11$, and the sum of their squares is 0.0708, which is smaller than can be obtained by any other values of x, y, and z. From (10) the probable error of each of the given observations may now be found to be $\pm 0''.13$.

When the weights are unequal the method of Art. 7 is to be followed. As an example, let the following be three angles measured at the station O:

$$\begin{aligned} MOA &= 46^\circ\ 53'\ 29''.4 \quad \text{with weight } 4, \\ MOC &= 135\ \ 27\ \ 11\ .1 \quad \text{with weight } 9, \\ AOC &= 88\ \ 33\ \ 41\ .1 \quad \text{with weight } 2. \end{aligned}$$

Now let x and z be the most probable values of any two angles, say of MOA and MOC. Then the observation equations are

$$\begin{aligned} x &= 46^\circ\ 53'\ 29''.4, \quad \text{weight } 4, \\ z &= 135\ \ 27\ \ 11\ .1, \quad \text{weight } 9, \\ z - x &= 88\ \ 33\ \ 41\ .1, \quad \text{weight } 2. \end{aligned}$$

Next let x_1 and z_1 be the most probable corrections to the observed values of x and z, so that

$$\begin{aligned} x &= 46^\circ\ 53'\ 29''.4 + x_1, \\ z &= 135\ \ 27\ \ 11\ .1 + z_1, \end{aligned}$$

are assumed probable values of x and z. Let these be substituted in the observation equations, which thus reduce to

$$\begin{aligned} x_1 &= \ 0''.00, \quad \text{weight } 4, \\ z_1 &= \ 0\ .00, \quad \text{weight } 9, \\ x_1 - z_1 &= +\ 0.60, \quad \text{weight } 2. \end{aligned}$$

From these the normal equations are formed; they are

$$\begin{aligned} 6x_1 - \ \ 2z_1 &= +\ 1.20, \\ -\ 2x_1 + 11z_1 &= -\ 1.20, \end{aligned}$$

from which the most probable corrections are

$$x_1 = +\ 0''.2, \qquad z_1 = -\ 0''.1.$$

Finally, the adjusted values of the three angles are

$$\begin{aligned} x &= \ \ 46^\circ\ 53'\ 29''.6 = MOA, \\ z &= 135\ \ 27\ \ 11\ .0 = MOC, \\ z - x &= \ \ 88\ \ 33\ \ 41\ .4 = AOC. \end{aligned}$$

Here it is seen that the observation having the largest weight receives the least correction, which should of course be the case.

It is well to note that the numerical part of the assumed probable values may be anything that is convenient; thus in the last example $46^\circ\ 53'\ 00''.0 + x_1$ might be taken for x, and $135^\circ\ 27'\ 00''.0 + z_1$ for z, then the values for x_1 and z_1 would be found to be $+\ 29''.6$ and $+\ 11''.0$. The object of introducing x_1, y_1, and z_1 is, however, to make the numbers in the right-hand members of the observation and normal equations as small as possible, and this is generally secured by taking the corrections as additions to observed values.

After the adjustment is made the azimuths of all the lines radiating from the station are easily found by simple addition or subtraction, provided the azimuth of one line is known. Thus for the last example let the azimuth of OM be given as $279^\circ\ 04'\ 18''.4$, then the azimuth of OA is $325^\circ\ 57'\ 48''.0$, and the azimuth of OC is $54^\circ\ 31'\ 29''.4$.

Prob. 16. Angles measured at the station O between the stations D, K, M, and C gave the following results:

$$\begin{aligned} DOK &= 66^\circ\ 32'\ 43''.70, &\text{weight } 2,\\ KOM &= 66\ \ 14\ \ 22\ .10, &\text{weight } 2,\\ KOC &= 108\ \ 02\ \ 29\ .62, &\text{weight } 1,\\ MOC &= 41\ \ 48\ \ 07\ .02, &\text{weight } 2,\\ COD &= 185\ \ 24\ \ 47\ .65, &\text{weight } 2. \end{aligned}$$

State the observation equations, form and solve the normal equations, find the adjusted angles, and show that the adjusted value of COD is $185^\circ\ 24'\ 47''.41$ with a probable error of $\pm\ 0''.20$.

17. Errors in a Triangle.

The simplest triangulation is a single triangle in which one side and the three angles are measured in order to find the lengths of the other sides. The precision of the values found for these sides will depend upon the probable error of the base and the probable errors of the measured angles. The best triangle is one whose angles are each about 60 degrees, and a triangle having one angle less than 30 degrees is not a good one, as will now be shown.

In a triangle whose sides are a, b, and c, let the angles A, B, and C and the side a be obtained by measurement. The sides b and c then are

$$b = a\frac{\sin B}{\sin A}, \qquad c = a\frac{\sin C}{\sin A}.$$

Now suppose each angle to have a probable error r; then by the use of (11)′ the probable errors in b and c are found to be

$$r_b = br\sqrt{\cot^2 A + \cot^2 B}, \quad r_c = cr\sqrt{\cot^2 A + \cot^2 C}. \quad (17)$$

If A, B, or C is a small angle its cotangent is large and accordingly r_b and r_c may be great. As far as b is concerned the smallest value of r_b will obtain when $A = B$, and as far as c is concerned the smallest probable error results when $A = C$; or the three angles should be equal and each be 60 degrees in order that the precision of b and c should be the same and each be as small as possible.

As a numerical example let $a = 1\,000$ feet, $A = 90°$, $B = 10°$, $C = 80°$, and let the probable error in each angle be 1′. Here by computation $b = 173.65$ feet, $c = 984.81$ feet, and then

$$r_b = 173.65 \times 5.67 \times r, \quad r_c = 984.81 \times 0.176 \times r.$$

The value of r to be used here is 1′ expressed in radians, or $r = 1/60 \times 57°.3 = 0.000291$. Accordingly the probable error of b is 0.29 feet and that of c is 0.06 feet, so that the computed values of b and c have a large degree of uncertainty. It will be noticed that b, which is opposite the small angle B, is liable to a far greater error than is c.

For a second example take the triangle in which $a = 1\,000$ feet, $A = 60°$, $B = 60°$, $C = 60°$, and let the probable error in each angle be 1′. Here $b = 1\,000$ feet, $c = 1\,000$ feet, and $r = 0.000291$; then from (17) there is found $r_b = r_c =$ 0.24 feet, so that the probable error of the computed side b is less than in the previous case.

The uncertainty of a line is the ratio of its probable error to its length. Thus in the first numerical example the uncertainty of the computed value of b is $0.29/173.65 = \frac{1}{600}$ nearly, and that of the computed value of c is $0.06/984.81 = \frac{1}{16\,000}$ nearly. In the second example, however, the uncertainties of b and c are $0.24/1\,000 = \frac{1}{4\,200}$ nearly. An uncertainty of $\frac{1}{600}$ is greater than that of a rough linear measurement, and an uncertainty of $\frac{1}{10\,000}$ is greater than should occur in the lengths of the lines computed in precise triangulations. In primary geodetic triangulation work the uncertainty of the computed sides of the triangles is usually about $\frac{1}{300\,000}$; thus the probable error in a line 30 000 meters long would be 0.1 meters.

From formula (17) it is seen that the uncertainties in the computed values of b and c are

$$u_b = r\sqrt{\cot^2 A + \cot^2 B}, \quad u_c = r\sqrt{\cot^2 A + \cot^2 C}, \qquad (17)'$$

and hence these may be computed without knowing the lengths of the sides b and c. If the probable errors of A, B, and C are different, let them be represented by r_1, r_2, and r_3; then from (12),

$$u_b = \sqrt{r_1^2 \cot^2 A + r_2^2 \cot^2 B}, \quad u_c = \sqrt{r_1^2 \cot^2 A + r_3^2 \cot^2 C}, \quad (17)''$$

are the uncertainties in the computed lengths of b and c. If the base a has a probable error r_a, this may also be taken into account by (12), and it will be found that the term $(r_a/a)^2$ must be added to the other terms under the first radical in (17)''.

In laying out a triangulation it is not possible to locate the stations so that each angle may be approximately 60 degrees, but it should be kept in mind that this is the best possible arrangement and that it should be secured whenever feasible. Angles less than 30 degrees should not be used except in unusual cases, or when the distances computed from them are not to be used for the computation of other distances.

Prob. 17. In a triangle the adjusted values of the observed angles are 25° 18′ 07″, 64° 01′ 26″, and 90° 40′ 27″, each having a probable error of 1″. The length of the side opposite the smallest angle is 3 499.39 feet, and its uncertainty is $\frac{1}{50000}$. Find the uncertainties in the computed values of the other sides.

18. The Triangle Adjustment.

When the three angles of a plane triangle have been measured their sum should equal 180 degrees, but as this is rarely the case they are to be adjusted so as to fulfil this condition. This is readily done in any particular case by the methods of Arts. 6, 7, and 16, but more convenient rules for doing it will now be deduced.

First, let the three observed values be of equal weight, and let these be A, B, and C. Let x and y be the most probable values of A and B; then the observation equations are

$$x = A, \quad y = B, \quad 180^\circ - x - y = C.$$

Now let v_1 and v_2 be the most probable corrections to be applied to A and B in order to give the most probable values of x and y, or

$$x = A + v_1, \qquad y = B + v_2.$$

Substituting these in the observation equations, the latter reduce to

$$v_1 = 0, \quad v_2 = 0, \quad v_1 + v_2 = 180^\circ - A - B - C.$$

Letting d represent the small quantity $180^\circ - (A + B + C)$ the normal equations are found to be

$$2v_1 + v_2 = d, \qquad v_1 + 2v_2 = d,$$

whose solution gives $v_1 = \frac{1}{3}d$ and $v_2 = \frac{1}{3}d$, which are the corrections to be applied to A and B. Then the correction to be applied to C is also $\frac{1}{3}d$. Hence the rule: Subtract the sum of the angles from 180° and apply one third of the discrepancy to each of the measured values. For instance, if the three measured angles are 64° 12′ 19″.3, 80° 07′ 47″.0, and 35° 39′ 55″.8, their sum is 180° 00′ 02″.1, and the discrepancy d is − 02″.1. Then 0″.7 is to be subtracted from each angle, giving 64° 12′ 18″.6, 80° 07′ 46″.3, and 35° 39′ 55″.1 as the most probable values.

Secondly, let the three observed values be of unequal weight. Let these be A with weight p_1, B with weight p_2, and C with weight p_3. The observation equations are the same as before, but are weighted, namely,

$$\begin{aligned} v_1 &= 0, \quad \text{with weight } p_1, \\ v_2 &= 0, \quad \text{with weight } p_2, \\ v_1 + v_2 &= d, \quad \text{with weight } p_3. \end{aligned}$$

From these the normal equations are

$$\begin{aligned} (p_1 + p_3)v_1 + p_3v_2 &= p_3d, \\ p_3v_1 + (p_2 + p_3)v_2 &= p_3d, \end{aligned}$$

whose solution gives the corrections v_1 and v_2 and then the correction v_3 is $d - v_1 - v_2$. Accordingly the results are

$$v_1 = \frac{d}{p_1 P}, \quad v_2 = \frac{d}{p_2 P}, \quad v_3 = \frac{d}{p_3 P}, \qquad (18)$$

in which, for abbreviation, P represents $\frac{1}{p_1} + \frac{1}{p_2} + \frac{1}{p_3}$. These formulas show that the corrections are inversely as the weights, so that the angle having the smallest weight receives the largest correction. For example, let the weights of A, B, and C be 10, 5, and 1; then $v_1 = \frac{1}{13}d$, $v_2 = \frac{2}{13}d$, and $v_3 = \frac{10}{13}d$, so that the correction for C is ten times that for B and five times that for A.

If only two angles of a triangle are measured there can be no adjustment made. If A and B are given by observations these are the most probable values of those angles, and the most probable value of C is $180° - A - B$. In all precise primary work the third angle should be measured as a check, as also to show the precision of the observations, whenever it is practicable. Spires and other inaccessible points may, however, be used as stations in secondary triangulation.

Prob. 18. The observed angles of a triangle are 74° 19′ 14″.3 with weight 3, 35° 10′ 42″.6 with weight 7, and 70° 30′ 09″.4 with weight 9. Find the adjusted values of the angles.

19. Triangle Computations.

The computation of the sides of a triangle is a simple matter, one side having been measured as a base line or being known from preceding computations. The theorem used is that the sides are proportional to the sines of their opposite angles; thus in the triangle ABC let the side AB be known, then

$$\log CA = \log AB - \log \sin C + \log \sin B,$$
$$\log CB = \log AB - \log \sin C + \log \sin A.$$

In making these computations it is desirable that a uniform method should be followed, and the following form for arranging the numerical work is recommended, it being similar to that used by the U. S. Coast and Geodetic Survey.

COMPUTATION OF A PLANE TRIANGLE.

Lines and Stations.	Distances and Angles.	Logarithms.
AB	2753.53	3.4398898
C	49° 04′ 49″.28	0.1216914
A	90 21 24 .66	$\bar{1}$.9999916
B	40 33 46 .06	$\bar{1}$.8131011
CB	3643.95	3.5615728
CA	2369.64	3.3746823

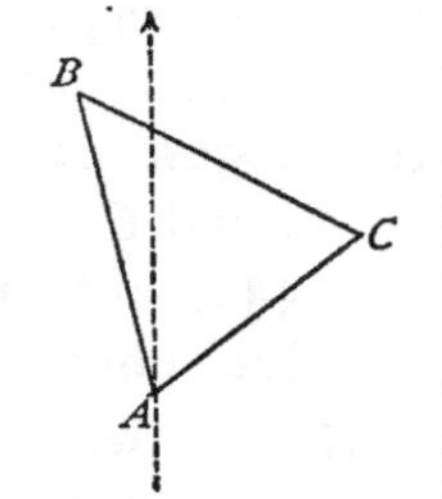

Here the stations are arranged in the order of azimuth, and that is placed first which is opposite to the given side, the length of this and its logarithm being put on the top line. Opposite the second and third angles are written their logarithmic sines, and opposite the first angle the arithmetical complement of its logarithmic sine. Now, to find the log of *CB* the logarithm opposite *B* is to be covered with a lead-pencil and the other three logarithms added. So to find the log of *CA* the logarithm opposite *A* is to be covered and the other three logarithms added. Lastly, the distances corresponding to these logarithms are taken from the table.

If the precision of angle work extends to seconds or tenths of seconds, as it does on primary triangulation, a seven-place table of logarithms will be needed. Six-place tables are rarely found conveniently arranged for rapid and accurate computation. For a large class of secondary work five-place tables are sufficiently precise. In taking a log sin from the tables the student should note that the characteristics 9. and 8. mean $\bar{1}$. and $\bar{2}$. and should write them in the latter manner in his computations.

When the above triangle ABC is connected with a coordinate system the azimuth of AB is known from previous computations. Then, from this and the angles A and B, the azimuths of AB and BC are easily found. Let the azimuth of AB be 149° 42′ 55″.68; then that of BA is 329° 42′ 55″.68, and accordingly

$$\text{Azimuth } AC = \text{azimuth } AB + \text{angle } A = 240° \; 04' \; 20''.34,$$
$$\text{Azimute } BC = \text{azimuth } BA - \text{angle } B = 289 \;\; 09 \;\; 09 \;.62.$$

As a check on these azimuths it may be noted that the second minus the first should be equal to the angle C.

The next computation is that of finding the coordinates of C from those of A and B. For the above triangle suppose that the coordinates of A have been assumed and that those of B have been computed from (13), the values being

Station.	*A*	*B*
Latitude	10 000.00	12 377.76
Longitude	8 000.00	9 388.59

and let it be required to compute the latitude and longitude of C. These should be found in two ways by the formulas in (13), so as to check the correctness of the results, and the form below shows how the numerical work may be arranged in a systematic manner. In the first column l denotes the length of AC or BC, the logarithm of the former being put in the third column and that of the latter in the fifth column. Similarly Z denotes the azimuth of AC or BC whose values are given in the second and fourth columns; adjacent to these are written the values of log cos$Z$ and log sin$Z$, $\bar{1}$. being written instead of the 9. in the tables. Then log l added to log cos$Z$ gives log l cos$Z$, and log l added to log sin$Z$ gives log l sin$Z$. The values of l cos$Z$ and l sin$Z$ are next taken from the logarithmic tables and placed in the second and fourth columns. Opposite L_1 and M_1 are placed the latitudes and longitudes of A and B, and the values of l cos$Z$ and l sin$Z$ are added to or subtracted from them as required by the signs of cos$Z$ and sin$Z$. It will be better, however, for the student to determine whether these are to be added or subtracted by drawing figures at the top of the table.

COMPUTATION OF COORDINATES.

Symbols.	C Computed from A.	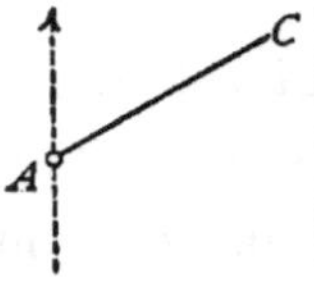	C Computed from B.	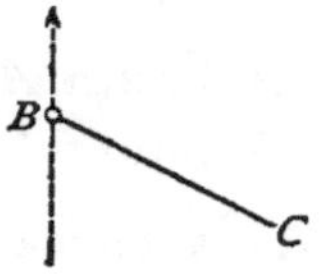
	Distances and Azimuths.	Logarithms.	Distances and Azimuths.	Logarithms.
l		3.3746823		3.5615728
	log cos$Z$ =	$\bar{1}$.6980292	log cos$Z$ =	$\bar{1}$.5159884
Z	240° 04′ 20″.34		289° 09′ 09″.62	
	log sin$Z$ =	$\bar{1}$.9378462	log sin$Z$ =	$\bar{1}$.9752699
l cos$Z$	1 182.26	3.0727115	1 195.53	3.0775612
l sin$Z$	2 053.66	3.3125285	3 442.25	3.5368427
L_1	10 000.00		12 377.76	
M_1	8 000.00		9 388.59	
lat. of C	11 182.26		11 182.23	
long. of C	5 946.34		5 946.24	

If the computations be correctly made the two values of the latitude of C must exactly agree, as also the two values of the longitude of C. In this case there is a discrepancy of 0.03 in the latitudes and of 0.10 in the longitudes and hence the numerical work must be revised so as to detect and remove the errors of computation.

Prob. 19. Revise all the computations in this Article and find the correct values of the coordinates of C. Also make the computations for the triangle DEF, in which $F = 95° 24' 01''.0$, $E = 54° 58' 08''.6$, $D = 29° 37' 50''.4$, $DE = 6\ 584.20$ feet, lat. $D = +15\ 328.75$ feet, long. $D = +12\ 047.05$ feet, azimuth $DE = 216° 17' 05''.6$, and determine the coordinates of E; finding lastly the coordinates of F in two ways.

20. Two Connected Triangles.

Let two triangles ABC and CDA have the side AC in common and let all the angles be measured, the observations being as follows and all of equal weight:

$$\begin{aligned} A_2 &= 45^\circ\ 19'\ 07'', & C_2 &= 50^\circ\ 19'\ 37'', \\ A_1 &= 48\ \ 07\ \ 15\ , & C_1 &= 37\ \ 46\ \ 50\ , \\ A &= 93\ \ 26\ \ 28\ , & C &= 88\ \ 06\ \ 15\ , \\ B &= 81\ \ 33\ \ 18\ , & D &= 96\ \ 54\ \ 19\ . \end{aligned}$$

Here it is seen that the sum of A_1 and A_2 is 06″ less than A, that the sum of C_1 and C_2 is 12″ greater than C, that the sum of A_1, B, and C_2 is 10″ greater than 180°, and that the sum of A_2, C_1, and D is 06″ greater than 180°. It is required to find the most probable values of the angles which entirely remove these discrepancies.

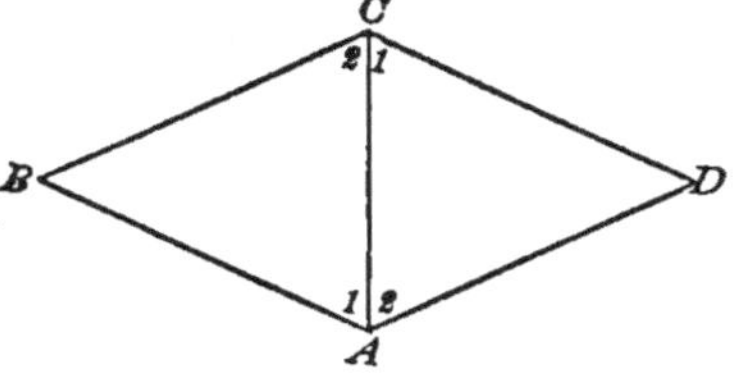

The number of observed angles is eight, but these are subject to the four conditions just mentioned, and accordingly there are really but $8 - 4 = 4$ independent angles to be used in the computation. Take A_1, A_2, C_1, and C_2 as these independent angles and let a_1, a_2, c_1, and c_2 be the most probable corrections to be applied to the observed values. The observation equations then are

$$\begin{aligned} a_1 &= 00'', & c_1 &= 00'', \\ a_2 &= 00\ , & c_2 &= 00\ , \\ a_1 + a_2 &= +06\ , & c_1 + c_2 &= -12\ , \\ a_1 + c_2 &= -10\ , & a_2 + c_1 &= -06\ . \end{aligned}$$

From these the normal equations are formed; they are

$$\begin{aligned} 3a_1 + a_2 + c_2 &= -04'', \\ a_1 + 3a_2 + c_1 &= 00\ , \\ a_2 + 3c_1 + c_2 &= -18\ , \\ a_1 + c_1 + 3c_2 &= -22\ , \end{aligned}$$

and the solution of these gives

$$a_1 = +00''.1, \quad a_2 = +01''.5, \quad c_1 = -04''.5, \quad c_2 = -05''.9,$$

as the most probable values of the corrections to the four angles. Then from the geometric conditions the corrections to the other angles are

$$a = a_1 + a_2 - 06'' = -4''.4, \quad c = c_1 + c_2 + 12'' = +1.6,$$
$$b = -a_2 - c_1 - 10'' = -4\ .2, \quad d = -a_1 - c_2 - 06'' = -3.0,$$

and applying these to the observed values they become

$$A_2 = 45^\circ\ 19'\ 08''.5, \qquad C_2 = 50^\circ\ 19'\ 31''.1,$$
$$A_1 = 48\ \ 07\ \ 15\ .1, \qquad C_1 = 37\ \ 46\ \ 45\ .5,$$
$$A\ = 93\ \ 26\ \ 23\ .6, \qquad C\ = 88\ \ 06\ \ 16\ .6,$$
$$B\ = 81\ \ 33\ \ 13\ .8, \qquad D\ = 96\ \ 54\ \ 06\ .0,$$

which are the most probable values of the angles and which at the same time satisfy the geometry of the figure.

When the observations are of unequal weights these are to be used in forming the normal equations from the observation equations. If one or more angles are unmeasured these do not appear in the observation equations and their values are to be derived from the adjusted results. If the angles A and C are not measured, but all the others are, then the only adjustment required is that of each triangle by the method of Art. 18.

If the length and azimuth of AB and the coordinates of A be given, the lengths and azimuths of the other lines of the figure, as also the coordinates of B, C, and D, may be computed by the methods of Art. 19. Thus a simple triangulation is established. When more than two triangles are connected the station adjustments are usually made first, and afterwards the triangle adjustments ; cases of this kind are discussed in Chapter IX.

Prob. 20. In the above figure let the observed values be as given except that of D, which is not measured. Find the adjusted values of all the angles.

21. Direct Observations with One Condition.

In Art. 18 are given examples where direct observations on several quantities are connected by a single conditional equation, and as other cases are to be discussed in future Articles it will be well to derive a general method of procedure which will simplify the numerical work. Let x and y be two quantities whose values have been found by observation, these having the weights p_1 and p_2. Let these quantities be connected by the conditional equation

$$q_1x + q_2y = D,$$

in which q_1 and q_2 are known coefficients, and D is a known quantity. Let v_1 and v_2 be the most probable corrections to the observed values so that the observation equations are

$$v_1 = 0, \text{ weight } p_1; \quad v_2 = 0, \text{ weight } p_2,$$

and the conditional equation reduces to

$$q_1v_1 + q_2v_2 = d.$$

Now let the value of one of these corrections be found from the last equation and be substituted in the observation equations, and then let the normal equations be formed and solved, and finally let the other correction be found from the conditional equation. The results will be

$$v_1 = \frac{q_1}{p_1}\frac{d}{P}, \qquad v_2 = \frac{q_2}{p_2}\frac{d}{P},$$

in which, for abbreviation, the letter P represents the quantity $\frac{q_1^2}{p_1} + \frac{q_2^2}{p_2}$.

The same process may be extended to any number of unknown quantities and similar formulas result. Thus if $v_1 = 0$, $v_2 = 0$, . . . $v_n = 0$, with weights p_1, p_2, . . . p_n, and if the conditional equation is

$$q_1v_1 + q_2v_2 + \ldots + q_nv_n = d,$$

then let $P = \frac{q_1^2}{p_1} + \frac{q_2^2}{p_2} + \ldots + \frac{q_n^2}{p_n}$, and the most probable values are

$$v_1 = \frac{q_1}{p_1}\frac{d}{P}, \quad v_2 = \frac{q_2}{p_2}\frac{d}{P}, \quad \ldots \quad v_n = \frac{q_n}{p_n}\frac{d}{P}, \qquad (21)$$

which also exactly satisfy the conditional equation. Formula (21) hence gives a general solution of this important case.

As a numerical example let there be measured at a station O the three angles $AOB = 97° \ 18' \ 20''$ with weight 5, $BOC = 135° \ 20' \ 05''$ with weight 3, and $127° \ 21' \ 29''$ with weight 6. Let x, y, and z be the most probable values, then must $x + y + z = 360°$. Take v_1, v_2, and v_3 as the corrections to the observations, and the conditional equation reduces to $v_1 + v_2 + v_3 = +06''$. Here $q_1 = q_2 = q_3 = 1$ and $d = 6''$; also $p_1 = 5$, $p_2 = 3$, $p_3 = 6$, and hence $P = 0.7$ and $d/P = +8.57$. Accordingly from (21) the values of the corrections are $v_1 = +1''.7$, $v_2 = +2''.9$, $v_3 = +1''.4$, so that the most probable values of the three angles which satisfy the conditional equation are $97° \ 18' \ 21''.7$, $135° \ 20' \ 07''.9$, and $127° \ 21' \ 30''.4$.

The above is the simplest application of the method of correlates which is extensively used in the adjustment of geodetic triangulations; further examples of it will be given in Chapter IX. For the case of equal weights the p's disappear from the above formulas and P becomes the sum of the squares of the q's. For instance, if the three observed angles of the last paragraph be of equal weight, then $P = 1 + 1 + 1 = 3$ and hence $v_1 = v_2 = v_3 = \frac{1}{3}d$, a result which agrees with the rule established in Art. 18; accordingly the adjusted values are found to be $19° \ 18' \ 22''$, $135° \ 20' \ 07''$, and $127° \ 21' \ 31''$, the sum of which is $180°$.

Prob. 21. The five interior angles of a pentagon, as found by measurement, are $80° \ 19'$, $120° \ 57'$, $107° \ 04'$, $141° \ 35'$, and $90° \ 00'$. Compute the adjusted angles, taking the weight of the last value as three times that of each of the others.

22. Intersections on a Secondary Station.

After a triangulation has been established any side may be used as a base from which to locate a secondary station by means of two measured angles. If, however, a third station is also used another computation may be made, and in general the results will not exactly agree with the first one owing to errors of observation. An adjustment is hence to be made in order to obtain the most probable position of the secondary station.

Let ABC be a triangle whose angles are known, it being a part of an established triangulation. At the three corners let the angles A_1, B_1, and C_1 be measured in order to locate a secondary station S. The lines determined by these angles do not in general meet at the same point, and hence the observations are to be adjusted to secure this result. The condition that the three lines shall meet in S is established by equating the expressions for the length of one as found from another in two ways; thus let BS be found, first by the triangle ABS and secondly through the triangles ASC and BSC; the values are

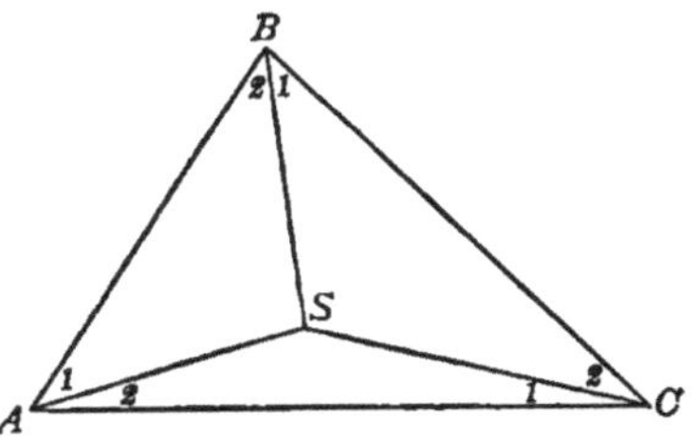

$$BS = AS\frac{\sin A_1}{\sin B_2}, \quad BS = AS\frac{\sin A_2 \sin C_2}{\sin C_1 \sin P_1},$$

and hence the conditional equation is

$$\sin A_1 \sin B_1 \sin C_1 = \sin A_2 \sin B_2 \sin C_2,$$

which must be exactly satisfied by the most probable values to be found for A_1, B_1, and C_1. This is called a side equation because it expresses the necessary relation between the three lines or sides which meet at S. By taking the logarithm of each member it becomes

$$\log\sin A_1+\log\sin B_1+\log\sin C_1-\log\sin A_2-\log\sin B_2-\log\sin C_2=0,$$

which is the form for practical numerical work.

As an example let the given angles of the triangle be $A = 83°\ 39'\ 01''$, $B = 57°\ 19'\ 42''$, and $C = 39°\ 01'\ 17''$. Let the three angles, as measured to locate S, be $A_1 = 41°\ 05'\ 10''$ with weight 2, $B_1 = 30°\ 15'\ 12''$ with weight 3, and $C_1 = 18°\ 46'\ 07''$ with weight 1; it is required to adjust these so that the three lines may meet in S and so that the values found may be the most probable.

Let a'', b'', and c'' be the corrections expressed in seconds to be added to the observed values A_1, B_1, and C_1. Then $41°\ 05'\ 10'' + a''$ is to be substituted for A_1 in the above conditional equation and similarly for B_1 and C_1. Now $\log\sin(A_1 + a'') = \log\sin A_1 + a''\cdot\text{diff. }1'$, where diff. 1′ is the tabular difference for one second corresponding to the value of A_1; thus $\log\sin(41°\ 05'\ 10'' + a'')$ is $\bar{1}.81769 + 0.23a''$, where 0.23 is in units of the fifth decimal place of the logarithm. In this manner the following tabulation is made:

Observed Angles.	Log. Sines.
$A_1 = 41°\ 05'\ 10''$	$\bar{1}.81769 + 0.28a''$
$B_1 = 30\ \ 15\ \ 12$	$\bar{1}.70228 + 0.35b''$
$C_1 = 18\ \ 46\ \ 07$	$\bar{1}.50751 + 0.61c''$
	$\bar{1}.02748 + 0.23a'' + 0.35b'' + 0.61c''$
$A_2 = 42\ \ 33\ \ 51$	$\bar{1}.83021 - 0.22a''$
$B_2 = 27\ \ 04\ \ 30$	$\bar{1}.65816 - 0.40b''$
$C_2 = 20\ \ 15\ \ 10$	$\bar{1}.53928 - 0.58c''$
	$\bar{1}.02765 - 0.22a'' - 0.40b'' - 0.58c''$

the values of A_2, B_2, and C_2 being those found by subtracting A_1, B_1, and C_1 from A, B, and C, and their corrections being the negatives of a'', b'', and c''. Equating the two members of the conditional equation, it reduces to

$$0.45a'' + 0.75b'' + 1.19c'' = 17,$$

while the observation equations are $a'' = 0$, $b'' = 0$, and $c'' = 0$, whose weights are 2, 3, and 1, respectively.

By the method of Art. 21 are now found $q_1^2/p_1 = 0.101$, $q_2^2/p_2 = 0.187$, $q_3^2/p_3 = 1.440$, and $P = 1.728$, whence $d/P = +9.83$. Then $a' = 0.225 \times 9.83 = +2''.2$, $b'' = +2''.4$, $c'' = +11''.8$, and finally the logarithmic corrections are $0.23a'' = +1$, $0.358'' = +1$, etc. Accordingly, the most probable values of the angles and of their logarithmic sines are found to be

Adjusted Angles.	Log. Sines.
$A_1 = 41°\ 05'\ 12''$	$\bar{1}.81770$
$B_1 = 30\ \ 15\ \ 14$	$\bar{1}.70229$
$C_1 = 18\ \ 46\ \ 19$	$\bar{1}.50758$
	$\bar{1}.02757$
$A_2 = 42\ \ 33\ \ 49$	$\bar{1}.83021$
$B_2 = 27\ \ 04\ \ 28$	$\bar{1}.65815$
$C_2 = 20\ \ 14\ \ 58$	$\bar{1}.53921$
	$\bar{1}.02757$

and these satisfy the geometric conditions of the figure as closely as can be done by the use of five-place logarithms. From these angles and the given lengths of AB, BC, and CA the distances AS, BS, and CS may now be computed.

The above method also applies when the point S is without the given triangle. Thus, if S be situated as shown in the figure, the above notation can be used by making $BAS = A_1$, $CAS = A_2$, $SCB = C_2$, and $SCA = C_1$. If the three points A_1, B_1, and C fall in the same straight line, the method fails, as then the conditional side equation is satisfied identically; in this case the distances AB and BC are known and a different side equation arises which involves these lengths.

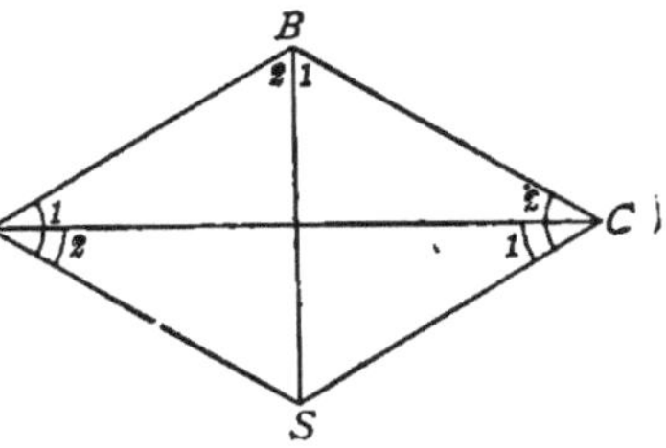

If in the last figure there be given the distances AB and BC and the angle B, and if A_1, B_1, and C_2 be observed, the condition that the lines AS, BS, and CS shall meet in one point is

$$AB \,.\, \sin A_1 \sin(B_1 + C_2) = BC \,.\, \sin C_2 \sin(A_1 + B_2),$$

which may be used in a manner similar to that of the above example. Thus let there be given $AB = 1\,067.950$ meters, $BC = 883.839$ meters, $B = 135° \ 50' \ 51''.6$, and let there be observed $A_1 = 75° \ 56' \ 00''.5$, $B_1 = 68° \ 34' \ 15''.2$, and $C_2 = 81° \ 06' \ 35''.0$, all of equal weight. Then by a similar process it will be found that the adjusted values of these angles are $A_1 = 75° \ 56' \ 08''.8$, $B_1 = 68° \ 34' \ 08''.6$, and $C_2 = 81° \ 06' \ 37''.0$, and that the two values of BC, computed from these, are equal.

Prob. 22. Let FG and GH be two parts of a straight line, each 800 feet long. At F, G, and H are measured the angles which lines from a station S make with the base, namely, $SFG = 40° \ 12'$, $FGS = 92° \ 58'$, and $GHS = 43° \ 55'$. Compute the length of GS in two ways, and, if they are not equal, find the most probable values of the angles which will effect an agreement.

23. The Three-Point Problem.

In secondary triangulation the position of a station S is sometimes determined by measuring the angles S_1 and S_2

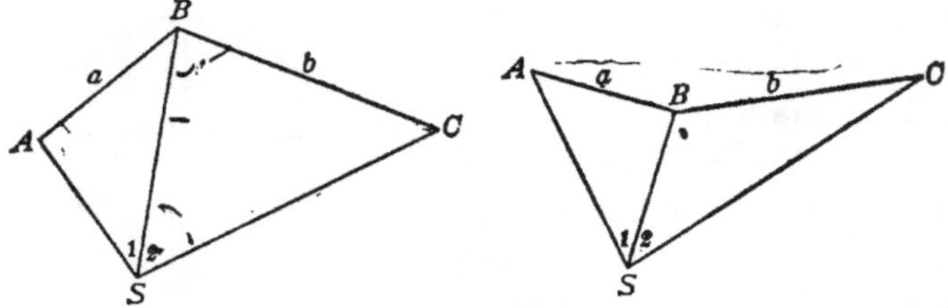

subtended at it by three stations A, B, and C whose positions are known. It is well to measure the three angles at S and then by the station adjustment find the most probable values of S_1 and S_2. The data of the three known points give the distances AB and BC which will be called a and b, and also the angle ABC which will be called B. The problem is to determine the distances SA, SB, and SC.

These distances can be found as soon as the angles A and C are known. Since the sum of the interior angles of the quadrilateral is 360 degrees,

$$A + C = 360^\circ - B - S_1 - S_2 \,;$$

and since the side BS is common to two triangles, the expressions for its length when equated give

$$\frac{\sin A}{\sin C} = \frac{b \sin S_1}{a \sin S_2}.$$

Thus two equations are established whose solution will give A and C. Let $A + C = 2m$ and $A - C = 2n$. The value of m is known, namely,

$$m = 180^\circ - \tfrac{1}{2}(B + S_1 + S_2), \qquad (23)$$

and that of n is to be found. Let V be such an angle that

$$\tan V = \frac{a \sin S_2}{b \sin S_1}; \qquad (23)'$$

then since $A = m + n$ and $B = m - n$, the second equation becomes

$$\frac{\sin(m + n)}{\sin(m - n)} = \cot V,$$

which is readily reduced to the form

$$\tan n = \tan m \cot(V + 45^\circ), \qquad (23)''$$

from which n is computed. The solution is hence made by first finding m from (23), secondly finding V from (23)', thirdly finding n from (23)'', and lastly the value of A is $m + n$ and that of C is $m - n$.

As a numerical example let the following be the given data for three stations, as determined by triangulation:

Line.	Azimuth.	Distance.	Station.	Latitude.	Longitude.
ID	327° 06′ 49″	9 011.0 ft.	*I*	34 104.2	52 581.5
DJ	74 56 58	5 794.5	*D*	26 537.2	47 688.9
JI	184 25 52	9 098.9	*J*	25 032.5	53 284.5

At a station S, within the triangle IDJ, there are measured the angles $ISD = 127° \ 47' \ 33''$, $DSJ = 87° \ 38' \ 18''$, and $JSI = 144° \ 34' \ 09''$. It is required to compute the lengths and azimuths of SI, SD, SJ, and also the coordinates of S.

Let station I correspond to A and station D to C; then drawing a figure and comparing it with that above, the data are $S_1 = 144° \ 34' \ 09''$, $S_2 = 87° \ 38' \ 18''$, $B = 74° \ 56' \ 58'' - 4° \ 25' \ 52'' = 70° \ 31' \ 06''$, $a = 9098.9$ feet, $b = 5794.5$ feet. Next $A + C = 57° \ 16' \ 27'' = 2m$, and $m = 28° \ 38' \ 14''$. From (23)′ log tan V is found, whence $V = 69° \ 43' \ 13''$, and then from (23)″ log tan n is found, whence $n = -14° \ 06' \ 42''$. Accordingly $A = 14° \ 31' \ 32'' = SIJ$, and $C = 42° \ 44' \ 56'' = JDS$. From the triangle ISJ are computed the distances $SI = 5600.6$ feet and $SJ = 3936.6$ feet; from the triangle JSD are found $SJ = 3936.6$ feet and $SD = 4417.3$ feet. The azimuth of SD is $74° \ 56' \ 58'' + 42° \ 44' \ 55'' + 180° = 297° \ 41' \ 54''$, and that of SI is $169° \ 54' \ 20''$. Lastly, the lengths of SI and SD are multiplied by the sines and cosines of their azimuths, giving the differences of latitude and longitude, which being added to or subtracted from the latitudes and longitudes of I and D furnish the coordinates of S in two ways. The latitude of S is found to be 28 590.4 feet and its longitude 51 600.0 feet.

A theoretic ambiguity is found in the above solution, since V and n may each have two different values corresponding to the values of tan V and tan n. This may be removed by always taking V as less than 90° and positive, and then taking n as less than 90° but making it positive or negative according as tan n is positive or negative.

When the point S in the above figure falls upon the circumference of a circle passing through P, Q, and R, the solution is indeterminate, as should be the case. When S lies very near this circumference the results of the computation will be uncertain. In such an event a fourth station should be used in the field work.

When more than three stations are observed from S there arises the N-point problem, in which three different locations for S can be computed by taking the stations three at a time. In this case a process of adjustment by the Method of Least Squares is to be followed so that the four lines may intersect in one point. This process will not be developd here, as it is of infrequent application and the numerical work is lengthy.

Prob. 23. Make the computations for the triangle IDJ from the above data, letting station I correspond to A and station J to C.

24. General Considerations.

A series of connected triangles with one or more measured bases may be called a triangle net. The purpose of the

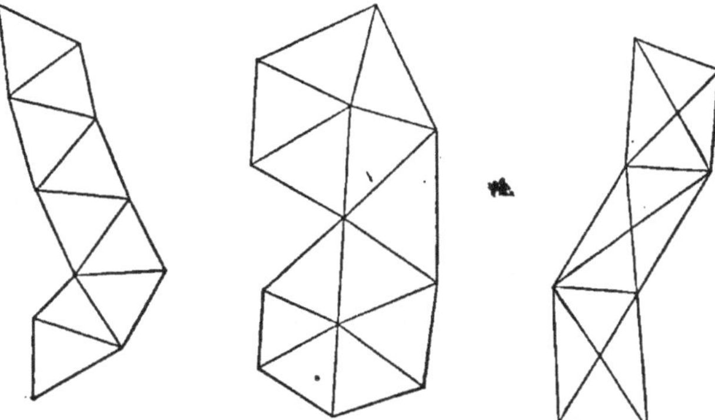

triangulation and the topography of the country will determine the location of the stations and the size of the triangles. A chain net is one suitable for a river survey, a polygonal net where the triangles from one or more polygons is sometimes used for a city survey, and a net composed of quadrilaterals each formed by four overlapping triangles is often used in geodetic work. The three types are, however, frequently combined together, single triangles, polygons, or quadrilaterals being used in different parts of the same net.

A chain net is the simplest in adjustment, since no side equation arises if there be but one base. In the other kinds there will be one side equation for each polygon and one for each quadrilateral, but increased labor in computation counts for little when precision is demanded. A quadrilateral is a figure securing high precision, and the polygon takes almost rank with it, since the side equation eliminates accidental errors that otherwise might be propagated along the net.

In the preceding pages only an introduction to the methods of adjustment has been given. The subject, however, will be continued in Chapter IX, where cases involving more than one conditional equation will be discussed.

It may have been noticed in the use of the side equation in the preceding Articles that the smallest angles receive the largest corrections if the weights of the observations are equal. It hence appears to be important in conducting the field work to measure angles less than 30° or greater than 150° with a higher degree of precision than those between 30° and 150°. By so doing the weight of the smaller angle will overbalance the error due to the large tabular difference in its sine, and the corrections will be more uniformly distributed among the measured values.

Geodetic triangulation nets differ from plane ones only in the greater size of the triangles and in the fact that the sum of the angles of each triangle is greater than 180°. All the preceding methods are hence directly applicable in geodetic work. When, however, a plane net is extended for some distance east or west of the meridian where the initial azimuth was determined, the computed azimuths become less or greater than the true ones owing to the curvature of the earth. In geodetic work this discrepancy is removed by introducing a correction which renders the back azimuth of a line different from its front azimuth, each value being the angle which the line makes with a meridian drawn through the end considered.

When the plane coordinates of two stations are known the length and azimuth of the line joining them is readily computed. Thus, let L_1 and L_2 be the given latitudes, then the latitude difference $L_2 - L_1$ is known; also let M_1 and M_2 be the given longitudes, then the longitude difference $M_2 - M_1$ is known. From (13) it is seen that the azimuth from the first point to the second is found by

$$\tan Z = -\frac{M_2 - M_1}{L_2 - L_1}, \tag{24}$$

and the distance l may then be computed from

$$l = \frac{M_2 - M_1}{\sin Z} \quad \text{or} \quad l = -\frac{L_2 - L_1}{\cos Z}. \tag{24$'$}$$

As an example, let the latitudes of two stations F and G be given as 15 420.72 feet and 18 115.13 feet, and their longitudes as 20 347.19 feet and 14 739.08 feet; here the latitude difference is + 2 694.41 feet and the longitude difference is

AZIMUTH AND DISTANCE FOR FG.

Symbols.	Distances and Azimuth.	Logarithms.	
$M_2 - M_1$	+ 5 608.11	3.7488165	
$L_2 - L_1$	− 2 694.41	3.4304637	
z	244° 20′ 17″.1	0.3183528	tan$z$
		$\bar{1}$.9549007	sin$z$
		$\bar{1}$.6365480	cos$z$
		3.7939158	
l	6221.80	3.7939157	

— 5 608.11 feet. The computation may be arranged in the form as shown. The second logarithm subtracted from the first gives log tan$Z$ and then Z is taken from the table; as tan$Z$ is positive 64° 20′ 17″.1 is the azimuth of GF and 244° 20′ 17″.1 is the azimuth of FG. Then log sin$Z$ and log cos$Z$ are taken out, and the subtraction of these from the first and second logarithms gives two values of log l which must agree within one unit of the last decimal. Lastly l is taken from the table. Thus the distance and azimuth between two stations which are not connected by a side of one of the triangles may be quickly computed in a plane system of coordinates.

Prob. 24. The latitudes of two stations M and N are 12 900.21 and 9 883.85 feet, and their longitudes are 27 333.16 and 35 640.93 feet. Compute the distance and azimuth from M to N.

Chapter III.

BASE LINES.

25. Principles and Methods.

The principle involved in the measurement of a base line is the same as that in common chaining, the unit of measure being applied successively from one end of the line to the other. It is very important that length of the measuring unit should be accurately known in terms of the standard linear foot or meter, for otherwise its absolute error may be multiplied so as to give an erroneous length for the base.

As the measuring bars or tapes are of metal they expand or contract as the temperature rises or falls and hence the coefficient of expansion of the metal must be known in order to eliminate errors due to this source. Other systematic errors, like those due to pull and sag in a tape and those due to the inclination of the base to the horizontal, must also be eliminated by computation. Accidental errors due to indefinite causes still remain in each result and, in order that the final length may be largely free from these, the measurement must be repeated several times and their mean be taken.

Metallic bars from 10 to 20 feet in length have been extensively used for base measurements. These are of two classes, end measures and line measures. With end measures the distance between the extremities of the ends is a unit, and measurement is made by contact, one bar being placed in position and another brought into line so that the ends of the two touch each other; these ends are usually rounded to a radius equal to the length of a bar. With line measures but one bar is required, the distance between two marks engraved upon its upper surface being a unit; a microscope being placed on a movable frame over one mark, the bar is moved forward until the other mark comes into the same position, and then the microscope is moved forward to the first mark. In each case the number of bar-lengths multiplied by the length of one gives the length of the base.

End measures are more convenient than line measures, but are generally not as precise. In order to eliminate effects of temperature, compound bars composed of metals whose rates of expansion are different have been devised and used; in these one bar expands more than another, so that by the use of a compensating lever the distance between the marks or ends is supposed to remain invariable.

Since 1885 the long steel tape has been extensively used in the measurement of base lines, and has been shown to give results of a high degree of precision. As such a tape can readily be bought and standardized, as its use involves little expert knowledge, and as a base can be measured with it very cheaply, a full explanation of the method of procedure will be given in later Articles.

Prob. 25. Consult Report of U. S. Coast Survey for 1897, and describe the duplex base apparatus, and ascertain the character of its work.

26. Probable Error and Uncertainty.

As a line is measured by the continued application of a unit of measure the probable error in a result found for its length should increase with that length. The law of this increase is found from formula (11); thus if r_1 be the probable error of the unit of measure and l be the length of the line, the probable error of l is

$$r = r_1 \sqrt{l}, \qquad (25)$$

that is, the probable error in a measurement of a line

increases with the square root of its length. Thus if two lines are measured with equal care and the second is four times as long as the first, the probable error of the second measurement is twice that of the first one.

Since weights are inversely as the squares of probable errors it follows that the weights of linear measurements made with equal care are inversely as the lengths of the lines. Thus, a measurement of 1 000 feet must be twice repeated and the mean of the results be taken in order to be worth as much as a single measurement of 500 feet. In combining linear measures, therefore, the weights of observations should be taken as the reciprocals of the distances.

The most convenient way to find the value of r_1 is to make duplicate measures of lines of different lengths. Let the lengths of the lines be $l_1, l_2, \ldots l_n$, the differences of the duplicate measures be $d_1, d_2, \ldots d_n$, and n be the number of lines. Then, as shown in treatises on the Method of Least Squares, the probable error of a linear unit is

$$r_1 = 0.4769\sqrt{\frac{\Sigma pd^2}{n}}. \qquad (26)$$

For example, in order to find the probable error of measurement with a steel tape four lines were measured as follows:

$l =$	427.34	854.21	1 281.71	1 708.40 feet
$l =$	427.37	854.20	1 281.74	1 708.33 feet
$d =$	− 0.03	+ 0.01	− 0.03	+ 0.07 feet
$p =$	0.00234	0.00117	0.00078	0.00059

Here the weights are taken as the reciprocals of the lengths, since the weight of a line one foot long is taken as unity. Then by the use of the formula the probable error of a measurement one foot long is found to be 0.00058 feet, and accordingly that of one 100 feet long would be 0.0058 feet. Of course a larger number of observations than four is required to deduce a reliable value of this probable error.

The uncertainty in the length of a line is expressed by the

ratio of its probable error to its length (Art. 26), and is hence given by $r_1/\sqrt{l}$, where r_1 is the probable error of a line one unit in length. Accordingly, if a certain line has an uncertainty of $\frac{1}{100\,000}$, the uncertainty of a line four times as long and measured in the same manner is $\frac{1}{50\,000}$. It thus follows that greater errors in the computed sides of triangles might result from a long base than from a shorter one.

Prob. 26. Let the probable error of measurement with a steel tape be 0.005 feet for 100 feet. A square city lot is laid out with this tape so as to contain 43 560 square feet. Show that the probable error of this area is 2.1 square feet.

27. Bases and Angles.

The uncertainty in the length of a computed side of a triangle is caused by a combination of the errors in the base with those in the angles, and the influence of the angles is usually greater than that of the base. Let the base a in the triangle ABC be measured with a probable error r_a, and let r be the probable error of the angle measurements expressed in radians. Then by (11),

$$r_b = b\sqrt{(r_a/a)^2 + r^2 \cot^2 A + r^2 \cot^2 B}$$

is the probable error in the computed value of b. Now in Art. 17 it was shown that the best-shaped triangle is an equilateral one, and for this case the formula gives

$$u_b = \sqrt{u_a^2 + \tfrac{2}{3}r^2}$$

as the uncertainty in the computed value of b. Let the probable error of the angle observations be one second or 0.000004848 radians. Then, if the base were without error, the uncertainty in b would be $\frac{1}{252\,600}$, but if the base have also an uncertainty of $\frac{1}{252\,600}$ the uncertainty in b will be $\frac{1}{178\,600}$.

It is not easy to carry on a triangulation so that the mean probable error of the adjusted angles shall be less than one second, but it is very easy to measure a base of moderate length so that its uncertainty shall be less than $\frac{1}{250\,000}$. In geodetic work bases have been measured with an uncertainty of less than $\frac{1}{1\,000\,000}$. It thus appears that even in the best-proportioned triangle the precision of the base measurement can be rendered greater than that of the angle work. The difficulty of finding good locations for bases and the expense of measuring them renders it customary, however, to use only one or two in a triangulation net of moderate extent. When the sides of the triangles are from one to ten miles in length a base line about a mile long may be used. Care must be taken that the triangles connecting it with the main net are well proportioned, no angle being less than 30 degrees. The topography of the country will determine the location of the stations to a great extent, but the figures show two methods

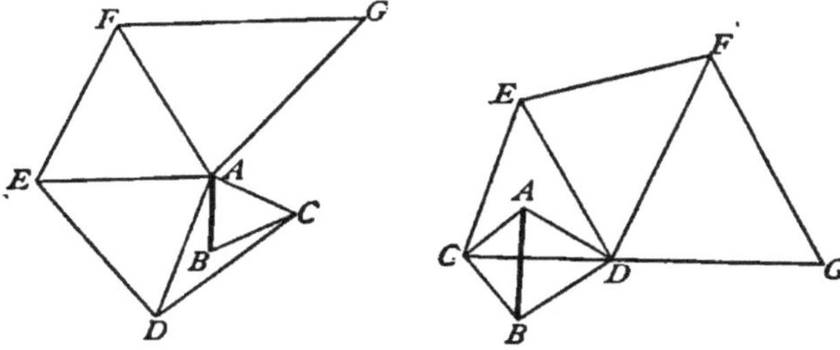

of gradually increasing the lengths of the sides away from a measured base AB; the second method is the better one.

In geodetic work bases several miles in length have been used. For example a base of the U. S. Coast and Geodetic Survey in Massachusetts is nearly 10¾ miles long, its measurement occupying three months in 1844. The final result, reduced to the ocean level, was 17 326.376 ± 0.036 meters, giving an uncertainty of $\frac{1}{481\,000}$. About 295 miles northeasterly is the Epping base, and 230 miles southwesterly is the Fire Island base, which were also measured with similar precision. The length of the Massachusetts base as computed through the triangulation from the Epping base was found to be 17 326.528 meters, and its length computed from the Fire Island base was found to be 17 326.445 meters. The actual uncertainties between the measured and computed values are hence $\frac{1}{114\,000}$ and $\frac{1}{250\,000}$ respectively, the effect of the errors in the angles being four times that of the base errors in the first case. In general it is found that angle errors do not increase the uncertainties of computed lengths to the extent that might be inferred from the preceding discussion, and this is probably due in part to the fact that they are largely eliminated in the adjustment of the triangulation.

Prob. 27. Four measures of a base line give the values 922.220, 922.197, 922.221, and 922.217 feet. Show that the uncertainty of the mean of these measures is about $\frac{1}{230\,000}$.

28. STANDARD TAPES.

A long steel tape is the most convenient apparatus for measuring the base line of a river or city survey, and it has also been used for geodetic bases with excellent results. It is necessary that it should be compared with a standard, and this can be done for a small fee by the Bureau of Weights and Measures at Washington. The certificate furnished will state the error of its length for a certain temperature and pull, or it will state that it is correct at a given temperature and pull. The coefficient of expansion, or the relative change in length for 1° Fahrenheit, should also be stated, in order that the effect of temperature may be eliminated. The coefficient of stretch, or the relative change in length for one pound of pull, must also be known. A tape thus standardized becomes itself a standard with which other tapes may be compared.

To compare another tape with the standard tape the coefficient of expansion of the latter must be known. To determine this the tape is stretched out on the floor of a large room whose temperature can be varied. With a spring balance at each end it is pulled to a certain tension, the thermometer noted and a certain length marked on two tin plates temporarily fastened on the floor. The temperature is then raised or lowered and the operation again repeated under the same pull. The change of length as marked on the tin plates is accurately measured, and this is divided by the total length and by the number of degrees to give the coefficient of expansion. The work should then be repeated several times using different lengths in each case, and the mean of the results be taken for the final coefficient.

If a tape is to be used under different tensions its coefficient of stretch should also be determined. The operation of doing this is similar to that above described, except that the temperature should be kept constant and the pull be varied. The change of length divided by the difference of the pulls and by the total length is the coefficient of stretch.

Sometimes a tape is stretched over two supports A and B, and thus owing to the sag the measured distance is too long. Let l be the distance read on the tape under a pull P, let d be deflection or sag at the middle, and w the weight of the tape per linear unit. The curve of the tape is closely that of a parabola, and if L be the horizontal distance, $L = l - \frac{8d^2}{3l}$ very nearly. Also taking moments about the middle of the span, $Pd = \frac{1}{2}wl \cdot \frac{1}{4}l$ nearly. Eliminating d from these two equations there results

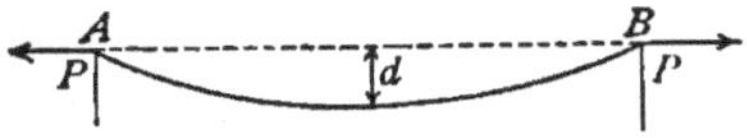

$$L = l - \left(\frac{1}{5}\frac{wl}{2P}\right)^2 l,$$

from which the true distance L can be computed from the observed distance l. If the distance AB be subdivided into

n equal parts by stakes whose tops are on the same level as those at A and B, then

$$L = l - \frac{1}{24}\left(\frac{wl}{nP}\right)^2 l$$

gives the horizontal distance between A and B.

It thus appears that any observation of a distance read on a steel tape may contain three systematic errors due to temperature, pull, and sag. Let t be the temperature and p the pull at which the tape is a standard, let T be the temperature and P the pull at which a measurement l is taken, let e be the coefficient of expansion, and s the coefficient of stretch, let w be the weight of the tape per linear unit, and if sag exists let n be the number of equal spaces in the distance l. Then the reading l is to be corrected by applying the following quantities:

$$\text{Correction for temperature} = + e(T - t)l,$$
$$\text{Correction for pull} = + s(P - p)l,$$
$$\text{Correction for sags} = - \frac{1}{24}\left(\frac{wl}{nP}\right)^2 l.$$

As an illustration, let $t = 56$ degrees, $p = 16$ pounds, $e = 0.00000703$, $s = 0.00001782$, $w = 0.0066$ pounds per linear foot. Let a horizontal distance 309.845 feet be read at a temperature of 49½ degrees under a pull of 20 pounds, there being 7 subdivisions in that distance. Then the correction for temperature is $-$ 0.0142 feet, that for pull is $+$ 0.0221, and that for sag is $-$ 0.0028 feet. The corrected measured distance is then 309.850 feet.

Lastly, if the measurement is made on a slope it must be reduced to the horizontal. For this purpose the difference of elevation of the two ends is found by leveling. Let h be this distance and L the length on the slope, then the horizontal distance is $L\sqrt{1 - \frac{h^2}{L^2}}$. For instance if the length

309.850 feet has 2.813 feet as the difference of level of the ends, then the horizontal distance is 309.838 feet.

Steel tapes used in base-line work usually vary in length from 300 to 500 feet. They have division marks at every 50 feet, but near the ends the marks are one foot apart, and a finely graduated rule is used for reading decimal parts of a foot.

Prob. 28. A tape is a standard at 41° F. when under 16 pounds pull and no sag, its coefficient of expansion being 0.0000069 and its coefficient of stretch 0.00000195. Find the pull P so that no corrections will be necessary when measurements are made at a temperature of 38 degrees and with no sags.

29. Measurement with a Tape.

When a base is to be measured with precision it should be laid out into divisions, each shorter than the length of the tape, and stout posts be set at its ends and at the points of division. In these posts are placed metallic plugs, each having drawn upon it a fine line at right angles to the direction of the base. The elevations of these plugs should be carefully determined by leveling.

Each division is then subdivided into several equal parts by light stakes set in line and on grade, the distance between the stakes being fifty feet or less. The tops of these stakes should be smooth and rounded so that friction may not prevent the transmission of a uniform tension throughout the tape; on the top of each stake two small nails may be driven to keep the tape in position. Instead of stakes special iron pins are sometimes used each having a hook to hold the tape.

The measurement should be done on a cloudy day with little wind in order to avoid errors due to change in temperature. The tape is suspended over two plugs and upon the intermediate stakes and pulled at both ends by spring balances to the desired tension. At one plug a graduation mark of the tape is made to coincide with the fine line on the plug, and at the other end the distance between the fine line and the nearest graduation mark is read by a closely graduated rule. Several measures of each division should be made at different times and with different pulls and the temperature be noted at each reading.

FIELD NOTES. BASE LINE *EG*. OCT. 3, 1888, P.M.

Divisions.	No. of Subdivisions.	Diff. in Elevation of Ends	Temperature.	Pull.	Measured Distance.	Remarks.
		feet	°	lbs.	feet	
III	7	2.813	51	16	309.865	
			50.5	18	309.857	
			50.5	20	309.842	
			50	16	309.870	
			50	18	309.857	Cloudy.
			49.5	20	309.845	
II	7	5.618	48	16	332.736	
			47.5	18	332.727	No Wind.
			47.5	20	332.712	
			47	16	332.740	
			47	18	332.726	
			47	20	332.715	
I	6	7.924	47	16	279.850	
			47	18	279.843	
			47	20	279.832	
			48	16	279.848	
			48.5	18	279.840	
			48	20	279.837	

The field notes of one measurement of a short base line *EG*, about 922 feet long, will illustrate the method of operation. There were three divisions, designated as I, II, and III, the first having six and the others seven subdivisions.

The steel tape used was about 400 feet long, and stated by its makers to be a standard at 56° Fahrenheit when under a pull of 16 pounds and having no sag. Its coefficient of expansion had been determined to be 0.00000703, its coefficient of stretch 0.00001782, and its weight per linear foot 0.0066 pounds. In order to correct the field results the expressions deduced in the last Article become

$$\text{Correction for temperature} = -\,0.00000703(56 - T)l;$$
$$\text{Correction for pull} = +\,0.00001782(P - 16)l;$$
$$\text{Correction for sag} = -\,0.000018815\frac{l^3}{n^2P^2};$$

from which the corrections are computed. For example, in division III, where $n = 7$, the mean of the observed distances is 309.856 feet and this is taken as the value of l in all the corrections. These being computed the corrected inclined distances are found and their mean gives 309.851 feet as the inclined length. Lastly, this is reduced to the horizontal, and 309.838 feet is the final length of division III.

COMPUTATIONS, DIVISION III, BASE *EG*.

Temp. T	Pull P	Measured Distance.	Corrections.			Corrected Distance.	Notes.
			Temp.	Pull.	Sag.		
°	lbs.	feet.	feet.	feet.	feet.	feet.	
51.	16	309.865	−.0109	−.0043	0	309.8498	$n = 7$
50.5	18	309.857	−.0120	−.0034	+.0110	309.8526	
50.	20	309.842	−.0120	−.0028	+.0221	309.8493	$h = 2.813$ ft.
50.5	16	309.870	−.0131	−.0043	0	309.8526	
50.	18	309.857	−.0131	−.0034	+.0110	309.8515	$C_g = -0.0128$ ft.
49.5	20	309.845	−.0142	−.0028	+.0221	309.8501	

Mean inclined distance = 309.851 ft.
Mean horizontal distance = 309.838 ft.

Proceeding in the same manner the corrections were found for Divisions I and II, and the sum of the three mean horizontal distances is 922.223 feet, which is the most probable length of the base line *EG* as determined from the observations of one day. Four other measurements of this base, made on four different days, gave the results 922.220, 922.221, 922.226, and 922.217 feet. The mean of these is 922.221 feet, whose probable error is 0.001 feet nearly, and accordingly the uncertainty of this final mean is about $\frac{1}{900\,000}$. It is thus seen that work of a high degree of precision can be done with a long steel tape whose constants are known.

The greatest errors in tape-line measurements are those due to errors in comparison with the standard and those due to the fact that the temperature of the metal is not the same as that of the air. The latter error may be removed by making some measurements when the temperature is rising and others when it is falling, and methods have also been devised of finding the exact temperature of the tape by means of an electric current passing through it; the former error cannot be removed except by the use of different tapes which have been independently compared with the official standard.

An account of the measurement of a geodetic base of 3780 meters, or about 2.3 miles, by steel tapes is given by Woodward in Transactions of American Society of Civil Engineers for October, 1893. It is concluded that the probable uncertainty in the final result, arising from all sources except that of error in the tape, cannot exceed $\frac{1}{2\,000\,000}$. This precision was secured by four days' work with twelve men, most of the measurements being made at night. In general it seems to be an established conclusion that precision in base measurements may be secured more cheaply by the use of tapes than by any other method.

Prob. 29. Correct the measurements on Division I of the above base line *EG*, and compute the most probable value of its final length and its probable error.

30. Broken Bases.

A base line should be perfectly straight and its ends be intervisible, but cases sometimes arise where obstructions, like a river or swampy land, render direct measurement impracticable. In geodetic work such a location should not be selected for a base line, but in secondary plane triangulation it may be used if expense is thereby avoided.

The first case is where the base AB is computed from two distances a and b, measured along the lines BC and AC. The three angles of the triangle are also measured and adjusted. The length of the base is then computed from $AB = b\cos A + a\cos B$, or from

$\overline{AB}^2 = (a + b)^2 - 4ab \sin^2\frac{1}{2}C.$

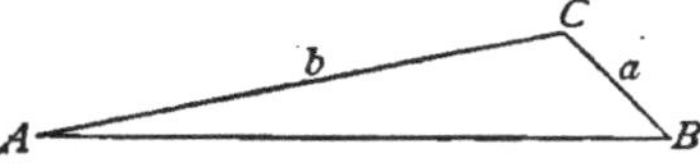

It might at first be thought that the small angles would introduce a high uncertainty in the computed length, but on reflection it is seen that this is not the case because two sides of the triangle are given, and accordingly the uncertainty due to the angles decreases with their sines. For instance, if $a = b$ and if $C = 170°$, it will be found that a probable error of one minute in C produces an uncertainty of only $\frac{1}{100\,000}$ in the computed length of the base.

A second case is where a stream crosses the base line between B and C. Here four points are selected on the line, two on each bank, and at these the angles are read which the base makes with lines drawn to an auxiliary station S. From these angles and the measured distances AB and CD the distance BC is computed in two ways, namely,

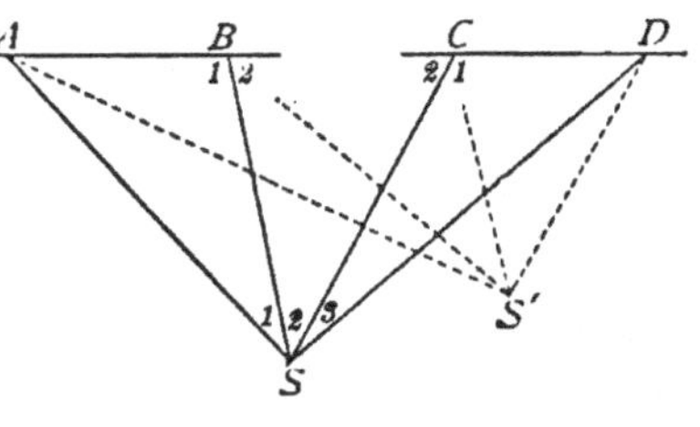

$$BC = AB\frac{\sin A \sin(B_2 + C_2)}{\sin C_2 \sin(A + B_1)} = CD\frac{\sin D \sin(B_2 + C_2)}{\sin B_2 \sin(D + C_1)},$$

and the angles should be measured with such precision that these agree in the last decimal used in the numerical work.

Another method of procedure in the last case is to measure only the angles at the station S. Let these be called S_1, S_2, and S_3, as shown in the figure, and let AB and CD be called a and b. Then the distance BC may be computed from

$$BC = \frac{a-b}{2\cos y} - \frac{a+b}{2},$$

where $a - b$ is to be taken as always positive and where y is an angle whose value is found from

$$\tan^2 y = \frac{4ab}{(a-b)^2} \frac{\sin(S_1 + S_2)\sin(S_2 + S_3)}{\sin S_1 \sin S_3}.$$

This is the method recommended by the U. S. Coast and Geodetic Survey. The demonstration of these formulas may be easily made by applying the second equation of Art. 23 to the three points ABC and then to BCD, and equating the two expressions each of which contains the unknown distance BC. In order to verify the result another station S' may be selected, the angles be measured there, and another computation be made.

Prob. 30. For the last case let there be given $a = 90.0242$ meters, $b = 120.0316$ meters, $S_1 = 19°\ 41'\ 44''.6$, $S_2 = 39°\ 20'\ 45''.2$, and $S_3 = 26°\ 19'\ 32''.8$. Using seven-place logarithms show that the length of BC is 107.8408 meters.

31. Reduction to Ocean Level.

Geodetic base lines must be reduced to mean ocean level in order that perfect agreement may obtain in the sides of triangles computed from different bases. Let AB be the base whose measured length is l and whose mean elevation above mean ocean level is h. Let ab represent this ocean level whose radius of curvature Ca or Cb is R. Then, from the two similar sectors, the value of *ab* is

$$l_1 = l - \frac{lh}{R}, \qquad (31)$$

and therefore the correction to be subtracted from the adjusted measured length is lh/R. For a long base this correction will be appreciable even when the base is but a few feet above the mean ocean level.

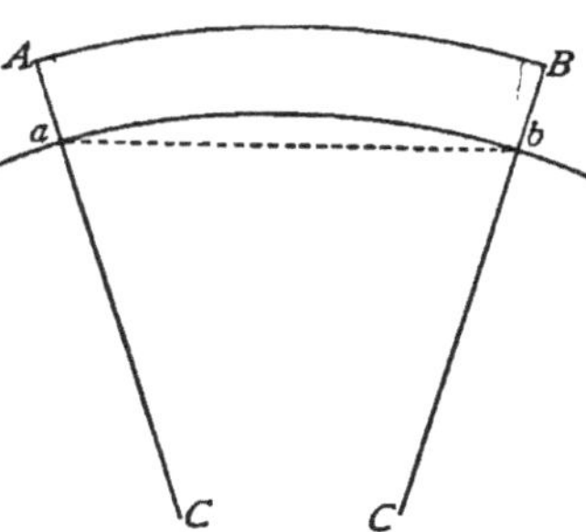

In Chapter VII it is shown how the radii of curvature have been found for different latitudes; it is there seen that for common cases the logarithm of R may be found by taking the mean of the logarithms of R_1 and R_2 given in Table IV at the end of this volume. When the azimuth of the base is given and great accuracy is required R should be computed from

$$\frac{1}{R} = \frac{\cos^2 Z}{R_1} + \frac{\sin^2 Z}{R_2},$$

in which Z is the azimuth of the base line and R_1 and R_2 are taken from Table IV.

For example, let the adjusted measured length of the base be 18 207.3267 meters, its mean height above ocean level 523.2 meters, and its mean latitude 40° 36′. From Table IV the logarithm of R is 6.8044705 and the correction lh/R is found to be 1.4943 meters, so that the length on the ocean level is 18 205.8324 meters. If the azimuth of the base be 75° 40′, the more accurate formula gives the logarithm of R as 6.8052175, from which the correction lh/R is 1.4917 meters, so that the final length on ocean level is 18 205.8350 meters.

It will be seen later that the triangles of a geodetic triangulation are computed by using the chords of the arcs of the spherical triangles. Accordingly the arc *ab* is to be reduced

to the straight line *ab*, which is the final length to be used for the base. As the curve is very flat, any one of the numerous approximate formulas for the relation between an arc of a circle and its chord may be employed, and

$$l_2 = l_1 - \frac{l_1^3}{24R^2} \qquad (31)'$$

is a convenient one for finding the correction to the arc l_1 by the use of logarithms. Thus for the above case, where $l_1 = 18\,205.8350$ meters and $\log R = 6.8052175$ meters, the correction to be subtracted from l_1 to give the chord l_2 is found to be 0.0062 meters, so that the length of the base to be used in triangle computations is 18 205.8288 meters, or, throwing off the last decimal place, 18 205.829 meters may be written as the final value.

Prob. 31. Compute the final length of the base line from the above data, supposing it be measured due north and south instead of in the direction stated.

CHAPTER IV.

LEVELING.

32. SPIRIT LEVELING.

The method of determining differences of elevation by an engineer's level and rod is called spirit leveling to distinguish it from the method in which vertical angles are used. In common work the telescope is made level by bringing the bubble into the middle of the attached scale. In geodetic work a sensitive bubble is used and readings of its ends taken on the scale, corrections to the rod readings being applied according to the distance of the rod from the instrument.

A level surface is one parallel to that of a fluid at rest, and the process of leveling consists in finding the elevations of points above the mean surface of the ocean. The line of collimation of the telescope of a properly adjusted and leveled

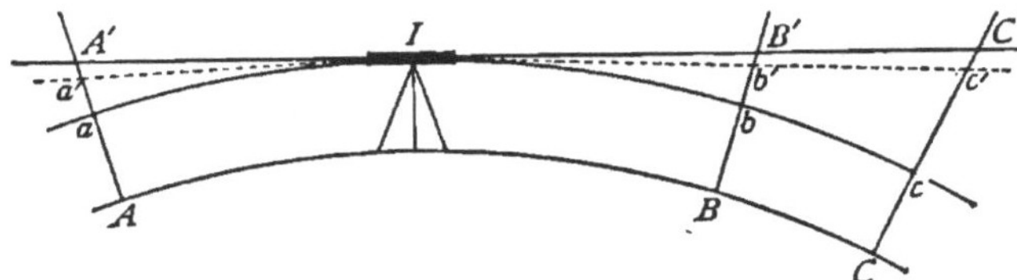

instrument, when revolved around the vertical axis, generates a plane which is tangent to a level surface. The line of sight, however, is depressed below that plane owing to refraction, and it lies between the tangent plane and the level surface, but nearer to the former. Thus if I be the telescope of the instrument, the straight line $A'B'$ represents the tangent plane, and the curved line ab the level surface, while the actual line of sight is $a'b'$, the points a' and b', in consequence of refraction, appearing to be in the tangent plane at A' and B'.

The rule that front and back sights should be of equal length in order to secure precision is one that is well known, and the figure shows the reasons for it. Let rods be set at A, B, and C in order to find the heights of B and C above A; then the observer will set the targets at a', b', and c', and the readings of the rods will be Aa', Bb', and Cc'. The height of B above A will be given by $Bb' - Aa'$, and that of C above A will be given by $Cc' - Aa'$. Now, owing to the combined effect of curvature of the level surface and of refraction of the air, the errors aa', bb', and cc' have been made in the rod readings, but the difference $Bb' - Aa'$ is the same as $Bb - Aa$ if the horizontal distances from B and A to the instrument are equal, while the difference $Cc' - Aa'$ is not the same as $Cc - Aa$ if the rod C is further from the instrument than the rod A.

It will be shown in Art. 37 that the deviation of a tangent plane from a level surface is about two-thirds of a foot at a distance of one mile and $\frac{2}{3}n^2$ feet at a distance of n miles, also that the deviation of the tangent plane from the refraction surface is one-seventh of that of the level surface. The combined effect of curvature and refraction is hence to cause an elevation of the line of sight above the level surface amounting to about 0.57 feet in one mile or $0.57n^2$ feet in n miles; a more exact rule is 0.000206 feet in 100 feet and $0.000206n^2$ feet in $100n$ feet. Thus, in the above figure, if the rods A and B be at 500 feet from the instrument, aa' and bb' are each 0.0051 feet, but the difference of level between A and B is free from error. If the rod at C be 1 000 feet from the instrument cc' is 0.0206 feet, and hence the difference $Cc' - Aa'$ is 0.0155 feet in error, for cc' is 0.0155 feet greater than aa'.

Another class of errors that is largely removed by taking backsights and foresights of equal length are those due to lack of perfect adjustment of the instrument. Thus if the line of collimation be not exactly parallel to the level bubble the reading on the rod at *A* may be too great, but when the sight is made on *B* the reading there is also too great, and hence these equal errors disappear in taking the difference of the rod readings. It is not desirable to try to do precise work with an instrument that is not in good adjustment, but it is essential to note that precise work cannot be done with back and front sights of unequal length, unless the lengths of these be measured and a correction be applied for the combined effect of curvature and refraction. In common work pacing, or even estimation, may be sufficient to prevent the introduction of these errors, but in precise work the distances should always be measured to the nearest foot.

Prob. 32. Let the rod readings at *A*, *B*, and *C* be 1.073, 3.137, and 9.271 feet, the distances from the instrument being 200, 250, and 400 feet. Find the elevations of *B* and *C* above *A*.

33. Duplicate Lines.

In common work with an engineer's level the precision of the elevations of the bench marks may be increased by running a second line between them and then taking the mean of the differences of level. This precaution can never be neglected in good work, for one measurement affords no data for estimating the precision of the results. It is better to run the two lines in opposite directions rather than in the same direction.

Semi-duplicate lines are those run in the same direction, having the same bench marks and heights of instruments but different turning points. Two sets of notes are kept which are not compared until a check is made on a bench mark. Thus in the figure let *M* and *N* be two bench marks and I_1, I_2, I_3, and I_4 the points where the level is set up, while A_1, A_2, and A_3 are the turning points on line *A*, and B_1,

B_2, and B_3 are the turning points on line B. The instrument being set at I_1 a backsight is taken on M and recorded in the

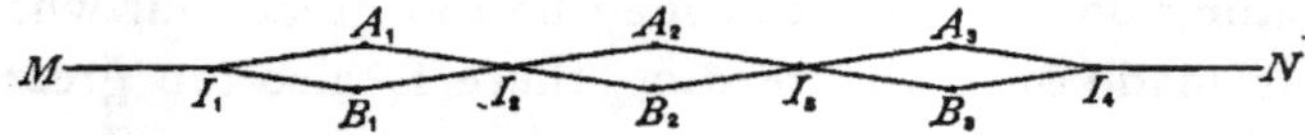

notes for line A; then another backsight is taken on M and recorded in the notes for line B. The two turning points A_1 and B_1 having been selected, foresights are taken upon them in succession and the readings recorded in the notes for lines A and B respectively. Then the instrument is moved to I_2 and backsights taken on A_1 and B_1 which are recorded in the separate notes for A and B. On arriving at I_4 backsights are taken upon A_3 and B_3 and two foresights upon N. Thus two lines $MA_1A_2A_3N$ and $MB_1B_2B_3N$ have been run between the bench marks M and N; if the elevation of N is to be determined from the given elevation of M, two sets of observations are at hand from whose comparison and combination it can be obtained with a higher degree of precision than by a single line.

Another method of running semi-duplicate lines is to have the same turning points but different heights of instruments. Thus, in the above figure, if I_1 be a bench mark the level is set at A_1, a backsight taken upon I_1 and a frontsight upon I_2; then the instrument is set at B_1, a backsight taken upon I_1 and a frontsight upon I_2. This method is not as convenient or expeditious as that above described, since it involves two rodmen, and it would be better to run two independent duplicate lines in opposite directions between the bench marks.

By taking proper precautions to preserve equality in the lengths of back and front sights, shading the instrument from the rays of the sun, and keeping the rod truly vertical, semi-duplicate lines may be run with an engineer's level so that the probable error of differences in elevation shall be less than 0.005 feet for bench marks one mile apart. In precise leveling where readings are taken to ten-thousandths of a foot, the probable error may be made much smaller. The adjusted elevations of the benches are of course the mean of the values found by the two lines.

It is sometimes observed that the elevations found by one line tend to be greater than those found by the other. For example, a line of semi-duplicate levels run from Bethlehem to Allentown, Pa., by students of Lehigh University in 1894 may be briefly noted. The total distance was 32 750 feet, this being divided into 27 sections with 28 bench marks. Computing the 27 differences of level for lines A and B it was found that nine were the same for both, that line A had nine greater and also nine less than line B; computing the elevations of the 27 benches from that of the Bethlehem bench it was found that 25 of these were greater on line B than on line A. The discrepancy between the two lines reached a maximum of 0.009 feet at 18 000 feet from the Bethlehem bench, then decreased to 0.001 feet, and afterwards increased until it became 0.005 feet at the Allentown bench. The probable error of the difference of level between the end benches, computed by the method of the next Article, was found to be 0.004 feet. This is perhaps a little smaller than would be found by independent duplicate lines run in opposite directions.

Prob. 33. The difference of level of two points P_1 and P_2 was found, by setting the level half-way between them, to be 6.438 feet. A second observation gave 6.436 feet, and a third one gave 6.437 feet. Show that the probable error of a single observation was 0.0007 feet.

34. Probable Errors and Weights.

The probable error of the difference in elevation of two bench marks increases with the number of times the instrument is set up between them, and will hence be greater in a hilly region than in a prairie country. It will also depend upon the precision of the instrument and upon the skill of the leveler and rodman, so that different classes of work will have different probable errors.

Assuming that the instrument is set up about the same number of times in a distance of one mile or one kilometer, it will be clear that the probable error in leveling is governed by the same law as that for linear measurements, namely that it increases as the square root of the distance. Thus if r_1 is the probable error in leveling a distance of unity, say one mile or one kilometer, then the probable error in leveling the distance l is $r = r_1 \sqrt{l}$. Thus if the probable error for a line one mile long is 0.006 feet the probable error for a line four miles long is 0.012 feet.

By means of duplicate lines of levels the probable error r_1 may be obtained by the application of formula (26), the weights being taken as the reciprocals of the lengths of the lines. Semi-duplicate lines, like those described in the last Article, may be used for the same purpose, but probably the value of r_1 found from them is somewhat smaller than from two lines run in opposite directions. As an example of the method, let D_a and D_b be the differences of elevation between two bench marks as determined by the two lines, d the differences, or discrepancies, between these, l the distance between the benches, and p the weight of d in terms of the weight of the unit of distance. Taking the following five measurements, and regarding 1000 feet as the unit of distance, the sum of the five values of pd^2 is 0.0000404

D_a	− 3.801	− 13.429	− 0.363	+ 5.528	+ 9.657 feet
D_b	− 3.803	− 13.426	− 0.365	+ 5.532	+ 9.653 feet
d	+ 0.002	− 0.003	+ 0.002	− 0.004	+ 0.004 feet
l	0.400	0.840	1.500	1.800	2.000 feet/1000
p	2.50	1.19	0.67	0.56	0.50
pd^2	0.0000100	0.0000107	0.0000027	0.0000090	0.000000080

and then from formula (26) the value of r_1 is found to be 0.0014 feet. Thus, for this class of work, the probable error in leveling a distance of 1 000 feet is 0.0014 feet, and hence the probable error in leveling any distance is $0.0014\sqrt{l}$, where l is the distance in thousands of feet. To find the probable error for one mile l is to be taken as 5.28, and thus $0.0032\sqrt{n}$ expresses the probable error of a line of levels n miles in length.

As weights are inversely proportional to the squares of probable errors it follows that the weights of differences of elevation are inversely proportional to the distances over which the leveling is extended. For example, let there be run three routes from P to Q giving the results

Route.	Miles.	P above Q.
1	5	37.407 feet
2	6	37.392 feet
3	10	37.414 feet

If the precision of the work per mile is the same, the value of r_1 being the same for the three lines, then the weights of the three results are to be taken as $\frac{1}{5}$, $\frac{1}{6}$, and $\frac{1}{10}$. The adjusted elevation of P above Q is then found by the rule of Art. 4 to be 37.403 feet.

The probable error r_1 may be also computed from lines run between two benches by different routes, as in the last example. The method to be followed is that of formula (9)″. Thus, taking the weight of a line one mile long as unity, the residuals v are found and the sum Σpv^2 is

	M	v	v^2	p	pv^2
	37.407	− 0.004	0.000016	0.20	0.00000320
	37.392	+ 0.011	0.000121	0.17	0.00002057
	37.414	− 0.011	0.000121	0.10	0.00001210
$z =$	37.403		0.000258		$0.00003587 = \Sigma pv^2$

formed. Then, from the formula, $r_1 = 0.0029$ feet, which is the probable error of a difference of level found from a line one mile long. Finally, the probable errors of the three observed differences of level are found from the square-root rule to be 0.0065, 0.0070, and 0.0092 feet, while the probable error of the adjusted elevation is 0.0039 feet, so that 37.403 ± 0.004 feet may be written as the final result.

Prob. 34. If the probable error in leveling one mile is 0.003 feet, what is the probable error in a line one kilometer long, and also in a line 100 kilometers long?

35. Adjustment of a Level Net.

When a closed circuit is made by running from *A* around to *A*, leaving the benches *B*, *C*, and *D*, the adjusted elevations of these are to be made by distributing the error of closure in direct proportion to the distances between the benches. For example, starting from *A* with the correct elevation of 420.317 feet above mean ocean level, the following elevations of other benches are found, and on returning to *A* its elevation is 420.467 feet, showing a discrepancy of 0.150 feet. The distances between the benches being 6, 3, 4, and 2 miles, $\frac{6}{15}$ of the discrepancy is to be subtracted from the elevation of *B*, $\frac{9}{15}$ from that of *C*, and so on. This method of adjustment is one that would

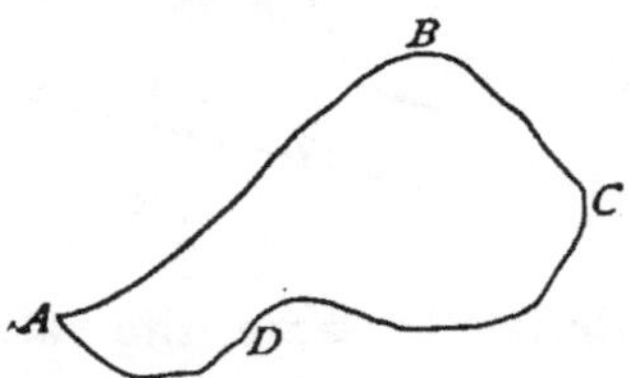

Bench.	Miles from *A*.	Observed Elevation.	Adjusted Elevation.	Correction.
A	0	420.317	420.317	0.000
B	6	532.918	532.858	− 0.060
C	9	607.200	607.110	− 0.090
D	13	510.315	510.185	− 0.130
A	15	420.467	420.317	− 0.150

be naturally used by every one, and it will be seen that it agrees with the results obtained by the application of the rule in Art. 21 to the determination of the most probable differences of the elevations between the benches.

A net of levels consists of several lines connecting benches in such a manner that the elevation of one can be deduced from another by several different routes. An example of the method of adjustment is given in Arts. 5 and 6, where, however, the weights of the different results are taken as equal. By introducing the weights according to the method of Art. 7, taking them as inversely proportional to the lengths of the lines, the same process may be applied to any given case. For example, take the case shown in the figure where eight differences of elevation between six benches are observed in a net consisting of three closed figures. These three figures give three geometric conditions and accordingly there can be but five independent quantities in the observation equations. This is perhaps seen more clearly by noting that, if the elevation of one bench be given, the elevations of the five others are to be obtained. In general the number of independent quantities in any net of level lines is one less than the number of benches.

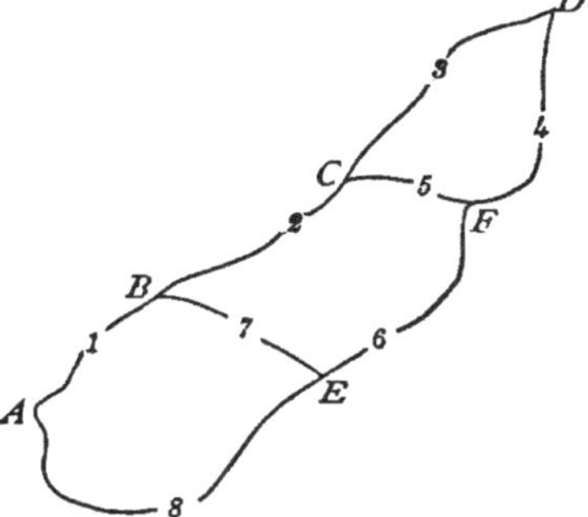

For example, let the eight observed differences of elevation be as given below, their weights being taken as the reciprocals of the distances between the benches. Let v_1, v_2, v_3, v_4,

No.	Benches.	Observed h. Feet.	Distance. Miles.	Weight.	Adjusted h. Feet.
1	B above A	12.02	4.0	0.25	12.039
2	C above B	23.06	7.2	0.14	23.012
3	D above C	14.30	5.0	0.20	14.340
4	D above F	29.44	6.3	0.16	29.389
5	C above F	15.02	2.0	0.50	15.049
6	F above E	9.34	4.8	0.21	9.372
7	B above E	1.45	3.5	0.29	1.410
8	E above A	10.67	8.3	0.12	10.630

and v_7 be corrections to be applied to the observed values h_1, h_2, h_3, h_5, and h_7 in order to give the most probable values. Then the eight observations are expressed in terms of these five quantities, the condition $h_3 - h_4 + h_5 = 0$ giving the fourth observation in terms of v_3 and v_4; thus,

$$
\begin{array}{llllllll}
1. & +v_1 & & & & & = & 0, \\
2. & & +v_2 & & & & = & 0, \\
3. & & & +v_3 & & & = & 0, \\
4. & & & +v_3 & +v_5 & & = & +0.12, \\
5. & & & & +v_5 & & = & 0, \\
6. & & +v_2 & & -v_5 & -v_7 & = & -0.15, \\
7. & & & & & +v_7 & = & 0, \\
8. & +v_1 & & & & -v_7 & = & +0.10.
\end{array}
$$

From these the normal equations are formed by the rule of Art. 7, using the given weights, and their solution furnishes the values $v_1 = +0.019$, $v_2 = -0.048$, etc., from which the above adjusted most probable values of the five quantities are found. Then the values of h_4, h_6, and h_8 immediately result.

The probable error for a line of levels one mile long can now be computed by formula (10). Each of the corrections being squared and multiplied by its weight, Σpv^2 is found to be 0.00246; then as $n = 8$ and $q = 5$, there results $r_1 =$ 0.019 feet as the probable error in leveling one mile, and accordingly $0.019\sqrt{l}$ is the probable error in leveling l miles. The degree of precision of the levels in this net is hence quite low compared with that required for city work.

Another method of stating observation equations in the above case is to take the elevations of five benches as the quantities to be found. Thus, if the elevation of A be given, approximate values of the elevations of the others are readily found, and the corrections to be applied to these may be called v_b, v_c, etc. Then each observation is expressed in terms of these corrections, and their most probable values are found by the solution of the resulting normal equations.

The adjusted elevations will be the same as those derived from the adjusted differences that are given above. Thus if the elevation of A be 312.724 feet, that of B will be 324.763 feet, that of E will be 323.353 feet, and so on.

Prob. 35. The elevation of a bench P is 725.038 feet. Level lines run between it and the benches Q, R, and S, give the following observations:

No.	Benches.	Difference in Elevation. Feet.	Distance. Miles.
1	B above A	35.080	3
2	C below A	8.698	6
3	D below A	19.905	4
4	C above D	11.212	3
5	C below B	43.780	3
6	B above D	54 995	6

State the observation equations, form and solve the normal equations, and find the adjusted elevations of the benches.

36. Geodetic Spirit Leveling.

Engineers' levels are of two types, the Y level and the dumpy level, the former being easier of adjustment while the latter is more precise. A dumpy level with two vertical and three horizontal wires in the diaphragm of its telescope, and having also a sensitive bubble, may be called a geodetic instrument. The rod is to be brought into the field between the two vertical wires, and readings taken upon it by each of the horizontal wires or by the help of a micrometer screw. The limits of the ends of the bubble are read upon the attached scale. The rod is provided with attached levels for securing verticality and it is set on a foot-plate planted in the ground. The distances from the instrument to the backsight and front-sight positions of the rod are measured.

Each instrument must be tested at intervals in order to determine the angular distance between the wires and the angular value of one division of the bubble scale. The usual adjustments for the level bubble and collimation axis are to be made, as also a series of measurements for determining the small errors still remaining in them. With these data tables can be made out for reducing the readings of each wire to the middle wire, for eliminating the error of inclination as determined by the readings of the ends of the bubble, and for eliminating the error of collimation.

As precise leveling for geodetic surveys is generally done in the metric system, the constants of Art. 32 are not directly applicable for the elimination of errors caused by unequal lengths of back and fore sights. If these distances be in meters and l_1 and l_2 their values, the former being the greater, then, for usual atmospheric conditions,

$$d = 0.0000675\ (l_1^2 - l_2^2) \qquad (36)$$

is the correction in millimeters to be subtracted from the difference in elevation $h_1 - h_2$. For instance, if $l_1 = 200$ meters and $l_2 = 170$ meters, then the difference $h_1 - h_2$ as found from the rod readings is 0.75 millimeters, or 0.00075 meters, too large. If the distances be in feet, then

$$d = 0.000206\ (l_1^2 - l_2^2) \qquad (36)'$$

is the correction in thousandths of a foot to be subtracted from the difference $h_1 - h_2$; thus if $l_1 = 656$ feet and $l_2 = 558$ feet, d is 2.5 thousandths of a foot or 0.0025 feet. These formulas are demonstrated in Art. 37.

The running of a line of geodetic levels is necessarily slow work, for daily tests of the instrument must be made and corrections applied to every rod reading in order to remove the errors above mentioned. The line is divided into sections from five to ten miles in length, each of which is leveled in opposite directions. The probable error of the elevations of the bench marks found by combining the two sets of observations has been made less than two millimeters for a distance of one kilometer, which is equivalent to about 0.008 feet for a distance of one mile.

Notwithstanding the apparent accuracy of leveling by one of the instruments above described, the item of cost is so high that it cannot be used except on government work. It may be remarked, further, that the probable errors deduced from the discussion of such level lines are but little, if any, less than those that can be obtained by good work under the common method. By rerunning the sections several times by the semi-duplicate plan of Art. 34, using a good engineer's dumpy level, and eliminating the systematic errors by making equal the lengths of backsights and foresights, it is not difficult to secure results whose probable errors shall be as low or lower than those of the so-called geodetic method, while the cost of the work per mile will be less than half as great.

Prob. 36. Consult Wilson's paper on "Spirit Leveling" in Transactions of American Society of Civil Engineers, 1898, Vol. XXXIX, pp. 339–430, and compare the methods used by the U. S. Geological Survey with those used by the U. S. Coast and Geodetic Survey. Also collect facts regarding the cost of running level lines.

37. Refraction and Curvature.

When light travels through air of varying density, its path is a curved line. If the surface of the earth were a plane a ray of light moving horizontally would suffer no refraction since the air would be of uniform density at all points in its path. Owing to the curvature of the earth a ray of light passing from c' to I, in the figure of Art. 32, travels through air of increasing density because c' is further than I from the level surface; similarly light passing from c to I tends to do so in a straight line, but encountering denser air its path becomes a curve which lies between the chord cI and the arc cI. Hence refraction is a consequence of curvature.

To develop formulas for the effect of curvature and refraction, it is necessary to take for granted that the earth is a globe whose mean radius R is about 3 959 miles or 6 371 kilometers. Let AO and BO be this radius in the exaggerated figure, AC a short distance l which is sensibly equal to the tangent AB, and bA the path in which light travels from b to A, thus making b appear at B to an observer at A. The deviation due to curvature in the distance l is hence represented by BC and that of refraction by Bb, their difference bC being the combined deviation; let these be called c, kc, and d respectively, k being an abstract number less than unity whose value will be shown later to be about $\frac{1}{7}$. Thus d is expressed by $(1 - k)c$.

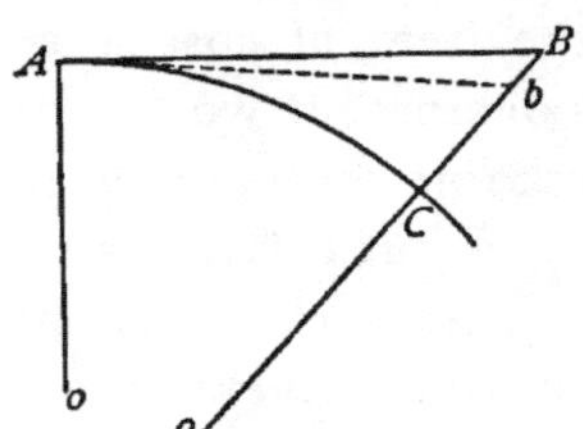

The value of c is readily found, from the right-angled triangle ABO, to be given by $2Rc + c^2 = l^2$, or since c is very small its square may be neglected, and thus

$$c = \frac{l^2}{2R}$$

is the deviation of the tangent plane from the level surface. The combined deviation due to curvature and refraction, or the distance bC, is then

$$d = (1 - k)\frac{l^2}{2R}. \qquad (37)$$

From this, using for k the mean value $\frac{1}{7}$, there is found

$$\left.\begin{aligned} d &= 0.0000000675l^2, \quad (d \text{ and } l \text{ in meters}), \\ d &= 0.0000000206l^2, \quad (d \text{ and } l \text{ in feet}), \end{aligned}\right\} \qquad (37)'$$

from which the formulas in Art. 38 directly result.

If the elevation of the eye of an observer above the ocean is known the distance to the sea horizon may be deduced from (38). Thus, for different systems of measures,

$$l \text{ (in kilometers)} = 3.85 \sqrt{d \text{ (in meters)}},$$

$$l \text{ (in statute miles)} = 1.32 \sqrt{d \text{ (in feet)}},$$

$$l \text{ (in nautical miles)} = 1.13 \sqrt{d \text{ (in feet)}}.$$

These results, like all in this Article, are mean ones, as curvature varies in different latitudes, and refraction varies under different atmospheric conditions.

Prob. 37. If the elevation d above the sea horizon is given in meters, what is the formula for l in nautical miles?

38. Vertical Angles.

The effect of refraction on any vertical angle is to render the measured value too large or too small according as it is an angle of elevation or angle of depression, while curvature produces the opposite effect. In the figure let A and B be two stations whose horizontal distance apart is l, the station B being higher than A. In order to find the difference in elevation the vertical angle of elevation BAC, or the vertical angle of depression ABD, is needed. Let an instrument be set at A and its horizontal limb be made tangent to the level surface AEC in the direction Ae; in consequence of refraction the station B appears to be in the direction Af, and fAe is the measured angle of elevation. The measured value is thus too large by the refraction angle fAB and too small by the curvature angle eAC; the true required angle BAC is hence $fAe - fAB + eAC$. In the same manner, when the instrument is set at B, the measured angle of depression is $f'Be'$, which is too small by the refraction angle ABf' and too large by the curvature angle DBe'. These effects of refraction and curvature are small, and sensibly the same at A and B under similar atmospheric conditions. Thus the combined effect of refraction and curvature renders the measured angle at A too small and that at B too large by the same number of seconds.

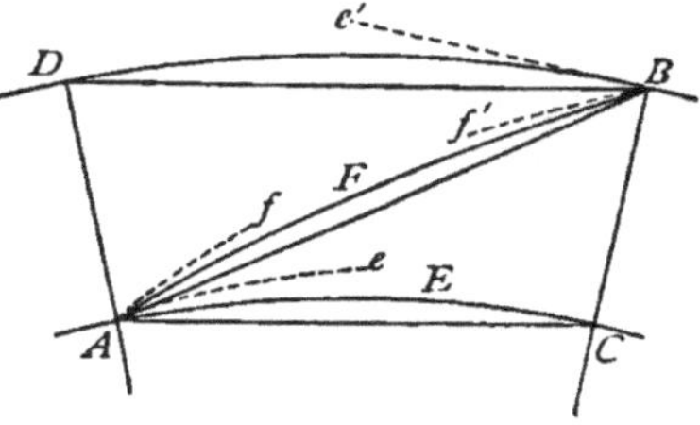

Let α be the angle of elevation at A and β the angle of depression at B, and let d'' be the correction in seconds, so that $BAC = \alpha + d''$ and $ABD = \beta - d''$ are the true required angles. In the last Article the linear correction d normal to l was found. The corresponding angle in radians is d/l and the corresponding value in seconds is $206\,265d/l$, and accordingly

$$d'' = 206\,265(1 - k)\frac{l}{2R} \qquad (38)$$

is the correction to be added to α and subtracted from β. Using the mean value $k = \frac{1}{7}$, and a mean value of R, there results

$$\left.\begin{array}{ll} d'' = 0.01394l, & \text{when } l \text{ is in meters,} \\ d'' = 0.00425l, & \text{when } l \text{ is in feet,} \end{array}\right\} \qquad (38)'$$

as the number of seconds to be added to angles of elevation and subtracted from angles of depression. Thus if the angle of elevation of a station 15 000 feet distant be observed to be 2° 19′ 07″ the true required vertical angle is about 2° 20′ 11″.

It is now to be shown how the coefficient of refraction k can be found. As the angles $\alpha + d''$ and $\beta - d''$ are equal the value of d'', if α and β are simultaneously measured, is $d'' = \frac{1}{2}(\beta - \alpha)$. Equating this to the former general value there is found for the coefficient of refraction

$$k = 1 - \frac{\beta - \alpha}{206\,265}\frac{R}{l},$$

in which $\beta - \alpha$ must be expressed in seconds. For instance let $\alpha = 2° \; 24' \; 58''.9$, $\beta = 2° \; 35' \; 34''.2$, and $l = 23\,661$ meters; then using for R the mean value 6 371 kilometers there is found $k = 0.165$. A better value of R is that of the radius of curvature of the level surface through the lower station A. Numerous simultaneous observations of vertical angles of elevation and depression have established that k varies from 0.12 to 0.18, a mean value frequently used being $k = 0.143 = \frac{1}{7}$. The average values deduced by the U. S. Coast and Geodetic Survey are $k = 0.158$ across parts of the sea near the coast, and $k = 0.130$ between primary triangulation stations at a high elevation.

If the elevation of the eye of an observer above the ocean is known the dip of the sea horizon in seconds may be expressed by combining the above value d'' with that of l given at the end of the last Article. Thus

$$d'' \text{ (in seconds)} = 58.8 \sqrt{d \text{ (in feet)}} = 106.5 \sqrt{d \text{ (in meters)}}.$$

Also if the angle of depression of the sea horizon be measured, its distance from the eye may be obtained from (39)′ and will be found to be 4.30 kilometers, 2.67 statute miles, or 2.32 nautical miles for each minute of vertical angle. These results are mean rough ones, since both curvature and refraction vary in different latitudes.

Vertical angles for determining heights are usually small, and hence a large probable error may occur in a computed height, even when the probable error of the vertical angle is not large. The formula $h = l \tan \alpha$ gives the height h in terms of the observed quantities l and α. Let l be supposed to be without error and let r be the probable error in α, then the probable error in h is $r dh/d\alpha$, or $lr/\cos^2\alpha$, which is practically lr, since $\cos \alpha$ is nearly unity. If r be expressed in seconds the corresponding probable error in h is $lr/206\,265$, or if r be expressed in minutes the corresponding probable error in h is $lr/3\,438$. Thus, if the probable error of a vertical angle be one minute, and the horizontal distance h be 6 876 feet, the probable error in the computed height h is 2 feet. In geodetic work, where leveling by this method is done between stations many miles apart, it is seen that the probable errors in the vertical angles must be rendered very low in order that the computed heights may have a fair degree of precision. The uncertainty in a computed height is $lr/206\,265h$, if r be in seconds; for example, if $l = 10\,352$ feet and $\alpha = 3°\,00'\,53'' \pm 03''$, then $h = 545.19 \pm 0.15$ feet, and the uncertainty of h is about $\frac{1}{3600}$.

Prob. 38. If the probable error of l be r_1 and that of α be r_2 show that the square of the probable error of h is found by $r_1^2 \tan^2 \alpha + r_2^2 l^2$, where the last term must be divided by $206\,265^2$ if r_2 is in seconds.

39. Leveling by Vertical Angles.

Leveling with the stadia and transit is often done in topographic work, and with care will give results whose probable error should be not greater than 0.5 feet in one mile or than $0.5\sqrt{n}$ feet in n miles. A greater degree of precision can be secured by measuring the horizontal distances with a tape, reading the vertical angles to half-minutes, selecting the stations so that the angles of depression are about equal in number to the angles of elevation, and having a fair uniformity in the lengths of backsights and foresights. In no case, however, can this work attain a degree of precision comparable with that done by spirit leveling.

The difference in elevation of two stations of a triangulation can be computed when the horizontal distance between them has been obtained. The best method is to make simultaneous observations of the angles of elevation and depression. Using the notation of the last Article, it is seen that $\alpha + d''$ and $\beta - d''$ are the true angles required, or since these values are equal the true vertical angle is expressed by $\frac{1}{2}(\alpha + \beta)$, and hence

$$h = l \tan\tfrac{1}{2}(\alpha + \beta)$$

is the required difference in elevation. It is thus seen that the effects of curvature and refraction are eliminated by taking the mean of the two observed vertical angles α and β. In this method it is, however, essential that the two measurements should be made as nearly simultaneously as possible in order that the same atmospheric conditions may affect both angles, for it is found that the coefficient of refraction varies with temperature and barometric pressure.

In a geodetic triangulation measurements of vertical angles are carried on at the same time with those of the horizontal angles, and it is not usually possible that the vertical angles at two stations can be simultaneously measured. Records of the weather are kept, however, and by taking at each station a considerable number of observations it is possible to select for any two stations several which are made under like atmospheric conditions. When this cannot be done values of the coefficient of refraction, determined for the region of the work, may be used, and the correction d'' to be applied to either angle may be found by (38)′. Then, either

$$h = l\tan(\alpha + d'') \quad \text{or} \quad h = l\tan(\beta - d'')$$

gives the difference in elevation of the two stations A and B.

The best time for measuring vertical angles is between 10 A.M. and 3 P.M., as between these hours the vertical refraction is less variable than either earlier or later in the day. The less the distance between the stations the less is the uncertainty in the refraction, and the larger the vertical angles the more reliable are the results. On account of the variability of refraction and the inherent inaccuracies of small angles, elevations found by vertical angles are far inferior in precision to those obtained by spirit leveling.

To illustrate the general method of procedure let A and B be the two station marks, whose horizontal distance apart is 10 352 feet. Let the instrument be set at A, the horizontal axis of the telescope being 6.1 feet above the station mark, and pointings be made on a signal b which is 27.5 feet above the station mark B. Let the mean of all

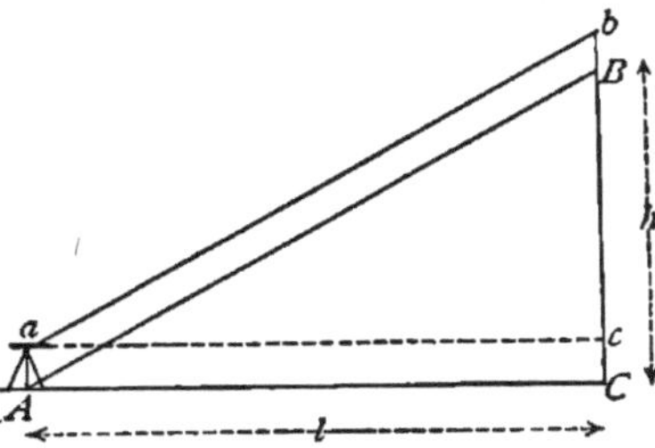

the observations give $3^\circ\ 07'\ 15''$ as the probable value of the angle of elevation bac. From (38)′ the mean correction for curvature and refraction is 44″, so that the vertical angle 3° 07′ 59″ is the value to be used. Then the difference of level between a and b is found by the use of the logarithmic tables to be 566.6 feet, and applying the correction for height of instru-

	Numbers.	Logarithms.	
l	= 10 352 feet	4.0150243	566.6
bac	= 3° 07′ 59″	$\bar{2}$.7382768	21.4
bc	= 566.6 feet	2.7533031	h = 545.2 feet

ment and signal the final difference in elevation between B and A is 545.2 feet. This is to be regarded as liable to a probable error of one foot or more on account of the uncertainties of refraction. To obtain a better result the angle of depression at B should be measured, another computation made, and the mean of the two results taken.

It is customary, when the distance l is large, to reduce the angle bac to BAC. Thus, if d represent the difference $Bb - Aa$, which is 21.4 feet in this case, the number of seconds to be subtracted from bac is $206\,265d/l$, or 426″. Then the angle BAC is 3° 00′ 53″, and l tan 3° 00′ 53″ gives

	Numbers.	Logarithms.
l	= 10 352 feet	4.0150243
BAC	= 3° 00′ 53″	$\bar{2}$.7215257
h	= 545.2 feet	2.7365500

at once 545.2 feet as the difference in elevation of the two station marks.

Prob. 39. The instrument is set 5.9 feet above B, pointing made on a signal 18.5 feet above A, and the angle of depression found to be 2° 57′ 30″. Compute the elevation of B above A.

CHAPTER V.

ASTRONOMICAL WORK.

40. FUNDAMENTAL NOTIONS.

In a triangulation covering an area of some extent it is desirable that the azimuth of one side should be determined by astronomical work in order that the computed azimuths may be all referred to the true meridian. Rough azimuths may be found by the magnetic needle or by making a noon-mark from shadows of a post cast by the sun. The method of obtaining the meridian with the solar compass or transit is known to all surveyors, and it gives results within about one or two minutes. Azimuth found with an engineer's transit from the sun or from Polaris furnishes results with about the same precision. For geodetic work, where an azimuth is desired with a probable error of only a few seconds, more accurate methods must be used.

In geodetic triangulations it is also necessary to obtain the astronomical latitude and longitude for a few of the stations, while those of the others are computed through the triangle nets. For the study of the figure of the earth these astronomical observations are especially important.

A brief outline of the field operations necessary for the determination of azimuth, latitude, and longitude will be presented in this Chapter. It is assumed that the student is acquainted with the fundamental notions regarding the circles of the celestial sphere, that he understands the method of locating the position of a star by its right ascension and declination, that he is familiar with the changes that occur throughout the year in the declination of the sun, and that he has a knowledge of the different ways of measuring time. In short, he should have had a good course in descriptive astronomy.

In geography the latitude of a place is its angular distance from the terrestrial equator, and in astronomy it is the angular distance of the zenith of the place from the celestial equator. Thus astronomical latitude is determined with reference to a vertical line at the point of observation. Since the horizon plane is perpendicular to this line it follows that latitude is the angular distance of the pole above the horizon.

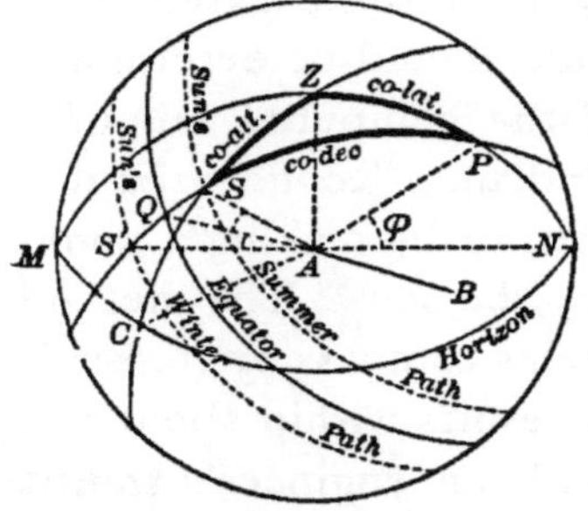

Thus, if A be any place, represented in the figure by a point at the center of the celestial sphere, the vertical line AZ determines the zenith Z. Let P be the celestial pole and N the north point of the horizon, then the angle PAN or the arc PN is the latitude of the place. In all questions relating to latitudes it is well for the student to remember that one minute corresponds approximately to one nautical mile on the earth's surface, and one second to about 101 feet (Art. 53).

The above figure sets forth several of the fundamental notions of astronomy. The horizon of the place A is shown by a circle joining N and C, the meridian of the place by the circle NPZ, and its co-latitude by the arc PZ. If S be the sun or a star QS is its declination and PS its co-declination, CS is its altitude and ZS its co-altitude or zenith distance. In the spherical triangle SZP the angle at P is the hour-angle of the sun or star, while the angle at Z gives its angular distance from the meridian, that is, its azimuth. In the summer the sun's apparent daily path lies north of the equator, at the equinoxes it lies on the equator, and in the winter it lies south of the equator.

Azimuths and hour-angles in astronomy are generally measured from the south around through the west from 0° to 360°, like azimuth in geodesy. It is, however, sometimes convenient to give them negative values; thus in the figure, if S represents the sun about 10 o'clock in the morning, its hour-angle ZPS is 330° or — 30°, and its azimuth MZS is about 310° or — 50°. Azimuths of circumpolar stars are sometimes estimated from the north toward the east and 180° is then to be added to give geodetic azimuths.

The methods to be here presented are those that can be carried out in the field with an engineer's transit or with a sextant. The results are not as precise as those derived with the instruments of an observatory or with portable astronomical instruments, but the fundamental principles and methods are the same and hence this Chapter may serve as an introduction to the practical field operations of geodetic astronomy. Azimuth is the most important problem for the civil engineer and it will be presented first, assuming that the latitude and longitude of the place have been found from a map and that standard time is given by a watch. Afterwards it will be shown how latitude, time, and longitude can be determined.

In most of the work of practical astronomy an almanac must be at hand to furnish the declinations, right ascensions, and other data that are needed in the computations. The American Nautical Almanac, which is published by the Bureau of Equipment of the U. S. Navy and can be had through any bookseller for fifty cents, will give all the data required for the work of this chapter.

Prob. 40. The declinations of the sun at Greenwich mean noon on March 20 and 21, 1900, were — 0° 13′ 28″.4 and + 0° 10′ 13″.4. When, in Eastern standard time, did the vernal equinox occur?

41. Azimuth by the Solar Transit.

With a transit having a solar attachment the azimuth of a line can be found by observing the sun at any time except between 11 A.M. and 1 P.M., the most favorable hours being generally from 9 to 10 A.M. and from 2 to 3 P.M. Such an attachment can be placed upon any transit at a cost of about fifty dollars. Accompanying it is a pamphlet giving full directions for use and adjustment, together with tables of the declination of the sun for Greenwich noon for each day of the year. Both the transit and the solar attachment should be in correct adjustment in order to do good work.

Let the upper part of the figure represent a section of the celestial sphere in the plane of the meridian, N and M being the north and south points of the horizon, P the pole, Z the zenith, Q the celestial equator, and S the place of the sun at noon. Let A be the point where the instrument is set, which may be regarded as the center of the celestial sphere. Then the angle QAZ, or its equal PAN, is the latitude of the place of observation. The angle QAS is the declination of the sun, which is positive when the sun is north of the equator and negative when it is south of the equator. The lower part of the figure is a plan, A being the place of the instrument, NM the true meridian, W and E the west and east directions, AS the direction of the sun about 10 o'clock in the morning, and AB a line whose azimuth is required to be found.

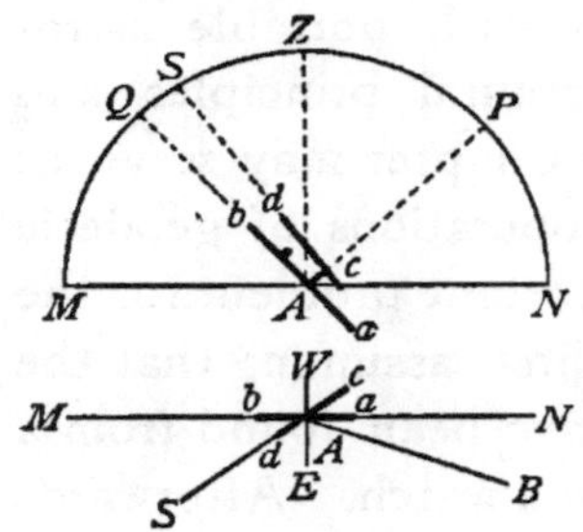

Let ab represent the telescope of the transit, it being represented as in the meridian and elevated so as to point to the celestial equator; this will be the case when the angle of elevation MAQ is equal to the co-latitude or when $MAQ = 90° - QAZ$. Let cd represent the telescope of the solar attachment pointing toward the sun; then the vertical angle between *ab* and *cd* is equal to the declination of the sun *QAS*. In this position the solar attachment is like an equatorial telescope, its axis pointing to the pole *P*, and as the sun moves the telescope *cd* can be made to follow it by simply turning it on its axis.

Before beginning the work a list of hourly declination settings is to be prepared by help of the table of declinations which is annually furnished by the maker of the instrument. This table also gives the corrections to be applied for refraction, these always being added to the true declinations, because refraction increases the true altitude of the sun. For example, let it be required to prepare the declination settings for the afternoon of September 16, 1899, for any place where Eastern standard time is used. The table gives $+ 2° 37' 44''.4$ as the declination of the sun at Greenwich mean noon of that day and $57''.91$ as the hourly decrease in declination. At 7 A.M. of Eastern standard time the declination is hence $+ 2° 37' 44''.4$, at 5 P.M. it is $+ 2° 37' 44''.4 - 10 \times 57''.91 = + 2° 28' 05''.3$, and at 4 P.M. it is $+ 2° 28' 05''.3 + 57''.9 = + 2° 29' 03''.2$. Thus the declination for each hour is found and placed in the second column. In

DECLINATIONS FOR SEPTEMBER 16, 1899.

Hour.	Declination.	Refraction Correction.	Declination Setting.	Remarks.
P.M.				
1	+ 2° 31′ 57″	+ 0′ 48″	+ 2° 32′ 45″	For Eastern Standard time.
2	+ 2 30 59	+ 0 54	+ 2 31 53	
3	+ 2 30 01	+ 1 05	+ 2 31 06	
4	+ 2 29 03	+ 1 32	+ 2 30 35	Lat. 40° 36′.
5	+ 2 28 05	+ 2 51	+ 2 30 56	

the third column are placed the refraction corrections as given in the table, and the fourth column contains the final declinations to be set off on the vertical arc as closely as its graduation will allow. The refraction correction is always additive, and hence if the declination is south or negative its numerical value is to be decreased, as the example for December 2, 1899, shows; for that day the table gives the declination at Greenwich mean noon as $-21° 58' 48''.3$ and the hourly change as $22''.20$.

DECLINATIONS FOR DECEMBER 2, 1899.

Hour.	Declination.	Refraction Correction.	Declination Setting.	Remarks.
A.M.				
8	− 21° 59′ 10″	+ 6′ 01″	− 21° 53′ 09″	For Eastern Standard time.
9	− 21 59 33	+ 2 59	− 21 56 34	
10	− 21 59 55	+ 2 11	− 21 57 44	Lat. 40° 36′.
11	− 22 00 17	+ 1 54	− 21 58 23	

After this list is made out the observer sets up the transit over the point A in order to find the azimuth of a line AB. The telescope is leveled by the attached bubble and pointed in a southerly direction. The declination setting for the hour is next laid off on the vertical arc, depressing the object glass if the declination is positive and elevating it if the declination is negative. The telescope of the solar attachment is then leveled by means of its own bubble, and thus the angle between the two telescopes is the same as the apparent declination, or the angle QAS in the above figure. Both telescopes are then elevated until the vertical arc reads an angle equal to the co-latitude of the place, or the angle MAQ. The solar attachment is next turned on its axis and the limb of the transit upon its axis until the sun is seen inscribed in the square formed by the four extreme cross-hairs in the focus of the solar telescope. When this is the case the transit telescope is in the plane of the meridian, and if desired a point may be set out in the line AS to mark that meridian.

It will be better, however, to read both verniers on the horizontal circle, then turn the alidade and sight on *B*, and read both verniers again. The angle *MAB* has thus been measured and, for the position in the figure, this is to be subtracted from 360° to give the geodetic azimuth of *AB*.

FIELD NOTES FOR AZIMUTH OF *AB*.

Time. October 28, 1895.	Reading on Meridian.				Reading on Line *AB*.				Angle *MAB*.			Remarks.
9:15 A.M.	20°	19′	00″	30″	182°	27″	30″	30″	162°	08′	15″	R. Doe,
9:30	80	00	15	15	242	08	30	30	162	09	00	Observer.
9:45	140	59	30	15	303	08	45	15	162	09	08	
3:15 P.M.	200	01	60	45	2	09	45	30	162	07	45	
3:30	260	12	45	30	62	22	15	30	162	09	45	$r_1 = 32''$
3:45	320	06	00	00	122	13	45	60	162	07	53	$r = 13''$
									Mean = 162° 08′ 38″			
									Azimuth of *AB* = 197° 51′ 22″			

The above form of field notes shows six observations made in this manner, and from their mean is found 197° 51′ 22″ for the azimuth of *AB*. The probable error of this mean is determined by Art. 9 to be about 13″, that of a single observation being 32″. This degree of precision is greater than can be generally attained by azimuth observations with the solar attachment, unless the observer has had considerable experience; nevertheless by a moderate amount of practice it is easy to determine an azimuth with a probable error of less than one-half a minute, both morning and afternoon observations being taken.

Prob. 41. Take several observations of the azimuth of a line by the solar transit, and find the probable error of their mean. Explain how the solar transit differs from the solar compass and state the advantages of the former over the latter.

42. AZIMUTH BY AN ALTITUDE OF THE SUN.

The azimuth of a given line may be determined by taking the altitude of the sun with an engineer's transit having a good vertical circle and reading the horizontal angle between the sun and the line. The latitude of the place must be known, and a nautical almanac must be at hand for finding the declination of the sun at the moment of observation.

In the figure let A represent the center of the celestial sphere, P the pole, Z the zenith, N the north point of the horizon, and S the position of the sun at the moment of observation. Then, in the spherical triangle PZS the angle Z is the azimuth of the sun measured from the north around through the east, and this is the same as the horizontal angle NAC. Let AB be the line whose azimuth is to be found; then if the horizontal angle CAB be measured its azimuth is known as soon as Z has been found.

In the figure CS is the altitude of the sun, and SZ is the complement of that altitude or the zenith distance of the sun; let the latter be represented by z. Let ϕ be the latitude of the place, or the arc NP. Let δ be the declination of the sun or the arc QS. Then in the spherical triangle PZS, the three sides are known, and hence the angle Z can be found from the equation of spherical trigonometry

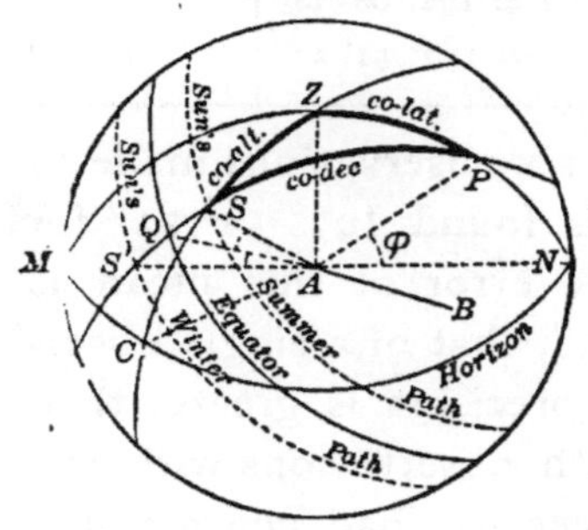

$$\sin\delta = \cos z \sin\phi + \sin z \cos\phi \cos Z.$$

For accuracy of computation, it is best to put this into another form; thus by writing $\cos Z = 1 - 2 \sin^2\frac{1}{2}Z$ a value is found for $\sin\frac{1}{2}Z$, and by writing $\cos Z = -1 + 2 \cos^2\frac{1}{2}Z$ a value is found for $\cos\frac{1}{2}Z$; then dividing the first value by the second there results

$$\tan\tfrac{1}{2}Z = \sqrt{\frac{\cos\frac{1}{2}(z + \phi + \delta)\ \sin\frac{1}{2}(z + \phi - \delta)}{\cos\frac{1}{2}(z - \phi - \delta)\ \sin\frac{1}{2}(z - \phi + \delta)}}, \quad (41)$$

from which Z is to be computed. In the figure S denotes the place of the sun in the summer half-year when δ is positive, and S' its place in the winter half-year when δ is negative. If the observation is taken in the forenoon the geodetic azimuth (Art. 13) of the sun is $180° + Z$, if in the afternoon it is $180° - Z$.

The transit having been put into thorough adjustment it is set up at A, the end of the line AB whose azimuth is to be determined. The horizontal limb being clamped, a reading of the horizontal circle is taken and the telescope pointed at B. The alidade is then unclamped and the telescope pointed at the sun, the objective and eye-piece being so focused that the shadow of the cross-wires and the image of the sun may be plainly seen upon a piece of white paper held behind the eyepiece. The cross-wires should be made tangent to the image on its lower and right-hand sides and the horizontal and vertical circles be read; next they should be made tangent to the image in its upper and left-hand sides and the two circles be read again. If the transit has a full vertical circle, which is necessary for the best work, observations should be taken both in the direct and reverse position of the telescope.

The following record will illustrate the method of making the measurements and obtaining the data for computation. The declination δ for 8:43 A.M., Eastern standard time, of the day of observation is taken from a nautical almanac. The mean apparent altitude is 43° 58′ 22″, and this being corrected for parallax and refraction the zenith distance z is found to be 46° 02′ 32″. By computation from the formula the azimuth Z, or the angle NAC, is found to be 101° 45′ 36″, whence the geodetic azimuth of the sun at the middle of the observation is 281° 45′ 36″. Subtracting from this the mean horizontal angle BAC the geodetic azimuth of the line AB is found to be 216° 44′ 06″.

AZIMUTH OF AB BY THE SUN.

Time. May 19, 1897.	Tel.	Vertical Angle. CAS.	Horizontal Angle. BAC.	Data and Results.
A.M.		Wires tan	gent to lower	ϕ = 40° 36′ 27″
		and right	sides.	δ at 7 A.M. = 19° 43′ 10″
				55
8h 40m	D	43° 09′ 00″	64° 48′ 00″	
42	R	43 35 30	65 10 30	δ = 19° 54′ 05″
		Wires tan	gent to upper	App. altitude = 43° 58′ 22″
		and left	sides.	Par.(+06″), Ref.(−60″) −54
8 44	R	44° 21′ 00″	64° 52′ 30″	True altitude 43 57 28
46		44 48 00	65 15 00	90
				z = 46° 02′ 32″
Means =		43° 58′ 22″	65° 01 30″	Z = 101° 45′ 36″
				180
				65 01 30
				Azimuth AB = 216° 44′ 06″

The correction for parallax of the sun is less than 9″ and it is always added to the apparent altitude. For an altitude of 20° the parallax correction is 8″, for 40° it is 7″, and for 60° it is 6″. In precise work the value of this correction may be found by multiplying 8″.9 by the cosine of the apparent altitude of the sun.

The refraction correction is taken from Table I at the end of this volume; it is always subtracted from the apparent altitude, since the effect of refraction is to render the apparent altitude greater than the true altitude.

The probable error of a single azimuth observation made by this method is usually one or two minutes; to secure a precise result several observations should be made both in the forenoon and afternoon and the mean of the computed values be taken. The best time for the work is when the sun is near the prime vertical, that is, nearly east or west. Near the noon hour the method is of no value, since then a small error in z causes a large error in Z; moreover when the sun is on the meridian $z = \phi - \delta$ identically.

More precise results can be obtained by using a star instead of the sun. In this case the observer looks at the image of the star in the telescope and brings it to coincide with the intersection of the cross-wires. The star is usually not bright enough to illuminate the cross-wires and hence it is necessary to throw light into the objective end of the telescope by means of a lamp held about a foot or two on one side of it. The signal at the end of the line AB must also be illuminated. The declination of the star is taken from the nautical almanac with less trouble than that of the sun as its daily change is inappreciable. The apparent altitude needs no correction for parallax, but the refraction correction is to be applied. With these exceptions the method of observation and computation is identical with that above explained. In the winter season, when the sun cannot be observed near the prime vertical, a star favorable for observation can always be found.

Prob. 42. In latitude 38° 53′ 18″, when the declination of a star was + 13° 55′ 33″, the apparent observed altitude was 28° 42′ 58″. Find the corrected zenith distance and compute the azimuth of the star.

43. Azimuth by Polaris at Elongation.

When Polaris is approaching its eastern or western elongation it may be easily followed by the vertical wire in the telescope of an engineer's transit, and when its motion in azimuth ceases a horizontal angle may be read between its direction and that of a given line. The azimuth of Polaris at elongation being known that of the line is immediately found.

In the figure let Z be the zenith, P the pole, N the north point of the horizon, HH the horizon itself, and E and W the positions of Polaris at the eastern and western elongations. *PN* is the latitude of the place of observation and hence *PZ* is the co-latitude $90° - \phi$; *PE* or *PW* is the co-declination of Polaris at elongation or $90° - \delta$; the angle *PZE* or *PZW* is the azimuth of Polaris at elongation measured eastward or westward from north. Let this azimuth be called *Z*; then, as the spherical triangles are right-angled at *E* and *W*,

$$\sin z = \cos\delta / \cos\phi, \qquad (43)$$

from which *Z* can be found for any given latitude when δ has been taken from the nautical almanac. The declination of Polaris is slowly increasing at the rate of about 19″ per year, its value being 88° 46′ 26″.6 for Jan. 1, 1900, and 88° 46′ 45″.4 for Jan. 1, 1901.

The approximate times of the elongations of Polaris for each month in the year are given in surveyor's handbooks and need not be repeated here. Half an hour before the time the observer sets the transit at *A* and places a signal, illuminated if necessary, at *B*. The horizontal circle is read, the lower limb being clamped, and the telescope is pointed at *B*; the alidade is then unclamped, the telescope pointed at Polaris, which is followed until it reaches its elongation, and then the horizontal circle is read again. Thus on August 15, 1899, at about 9:50 P.M. local time, an eastern elongation occurred and an observer in latitude 40° 36′ took the reading 87° 19′ 30″ when the pointing was made on *B* and 74° 04′ 00″ on Polaris; the horizontal angle *HAB* is hence 13° 05′ 30″. From the nautical almanac the value of δ is 88° 46′ 19″ and then by the formula the value of *Z* is 1° 37′ 04″. Thus for

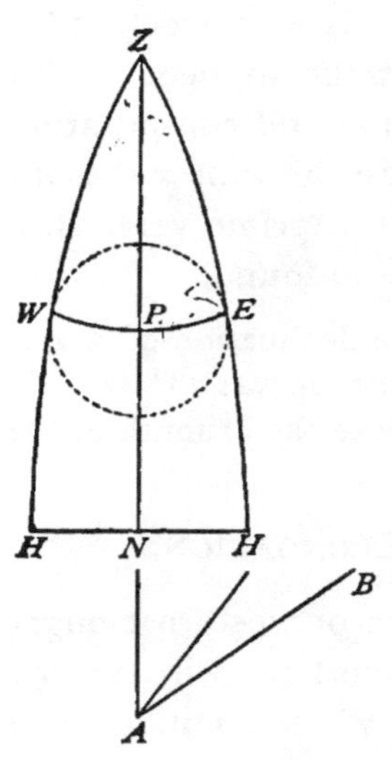

this case, as shown in the figure, the direction of AB is 14° 42′ 34″ to the eastward of the meridian and accordingly its geodetic azimuth is 194° 42′ 34″.

By the above method only one reading of the horizontal circle can be taken on Polaris, and hence there is no opportunity to eliminate the various sources of error of the transit. It is, however, possible to measure a number of angles before and after elongation and apply to each a correction to reduce it to elongation. For this purpose the time of elongation should be known and this can be found in local time within less than half a minute by the tables in the Handbook for Surveyors. Five pointings on Polaris may then be made during the quarter-hour preceding elongation and five during the quarter-hour following elongation. A good plan is to take these exactly at the beginning of three-minute intervals, then to read the verniers, turn to the signal or mark at B and read again. Half the angles are read with the telescope in the direct position and half with it in the reverse position. The readings are distributed over the circle by making each one about 20° greater than the preceding. The following form of field notes will render clear the method of conducting the work. The eastern elongation was to occur at 9:49 P.M. in the time indicated by the watch of the observer, and it was arranged to take the five pointings before elongation 14, 11, 8, 5, and 2 minutes earlier, while those following were taken 1, 4, 7, 10, and 13 minutes later, as shown in the second column. Thus ten horizontal angles were measured between the star and the illuminated mark at B. To reduce these to elongation a correction c is to be subtracted from each, this being computed from the formula $c = 0.07n^2$, where n is the number of minutes of time preceding or following elongation. These values of c are seen in the last column, and the mean corrected angle HAB is found to be 13° 05′ 33″, from which the final geodetic azimuth of AB is 194° 42′ 37″.

AZIMUTH BY POLARIS. AUGUST 16, 1899.

Time. P.M.	n	Tel.	Reading on Polaris.				Reading on Mark B.				Horizontal Angle.	c
	m		°	′	″	″	°	′	″	″	° ′ ″	″
9:35	+ 14	D	0	17	00	30	13	22	40	60	13 05 40	− 13
38	+ 11	R	20	42	30	50	33	48	45	50	13 05 47	− 8
41	+ 8	R	40	03	10	20	53	08	40	50	13 05 30	− 4
44	+ 5	D	60	40	40	50	73	46	00	10	13 05 20	− 2
47	+ 2	D	80	00	10	10	93	05	50	40	13 05 35	0
50	− 1	R	100	31	20	05	113	36	40	40	13 05 27	0
53	− 4	R	120	04	15	10	133	09	60	40	13 05 37	− 1
56	− 7	D	140	17	50	50	153	23	30	30	13 05 40	− 3
59	− 10	D	160	14	50	50	173	10	50	60	13 06 05	− 7
10:02	− 13	R	180	25	10	00	193	30	50	40	13 05 40	− 12

Time of elongation 9:49 P.M. Mean corrected $HAB =$ 13° 05′ 33″

$\delta =$ 88° 46′ 19″ $180 + Z =$ 181 37 04″

$\phi =$ 40 36 00 Azimuth of $AB =$ 194° 42′ 37″

$Z =$ 1 37 04 J. Doe, observer and computer.

For work with an engineer's transit the corrections can be found close enough south of latitude 50° by the approximate rule $c = 0.07n^2$, but for observations with a theodolite, where tenths of seconds are to be used, the more accurate formula given in treatises on practical astronomy should be employed.

The precision of this method depends almost wholly upon that of the pointings and readings. If the declination of Polaris be taken from the nautical almanac by interpolating for the day of observation no error can arise from this source. The error in the computed Z due to an error of one minute in the latitude will range from 1″ at latitude 30° to about 2″ at latitude 48°. The probable error of a single angle measurement may range from 5″ to 40″, depending upon the skill of the observer and the kind of transit or theodolite used, and accordingly the probable error of an azimuth found from a series of ten angles may range from 2″ to 15″. The precision of the series above given is considerably greater than can be generally secured by an engineer's transit reading to half-minutes.

This method is advantageous from its simplicity, but disadvantageous because at the utmost only two observations can be taken in twenty-four hours. For a single reading taken exactly at elongation the time need not be known further than to be sure of being ready a few minutes before it occurs. For several readings it should be known within half a minute, so that the times of the pointings may be arranged in advance.

Prob. 43. Show that the error in a computed azimuth due to an error in latitude increases with the tangent of the latitude; or if $d\phi$ is the error in latitude show that $dZ = -\tan z \tan\phi . d\phi$.

44. Azimuth by Polaris at any Hour-angle.

Polaris or any other circumpolar star may be used at any position for the determination of azimuth, if the observer's watch indicates correct time, either local or standard, and if the latitude and longitude of the place are known. From the time of observation and the data given in the nautical almanac the hour-angle of the star is to be found and the solution of a spherical triangle then gives the azimuth of the star.

Let Z be the zenith, P the pole, S the place of the star, and N the north point of the horizon. In the spherical triangle PZS the angle Z is the azimuth of the star east of the meridian, and the angle at P is its hour-angle minus 180°; the side ZP is the co-latitude $90° - \phi$, the side PS is the co-declination $90° - \delta$. Let t denote the hour-angle of the star, that is the obtuse spherical angle NPS; then the solution for the angle Z gives

$$\tan Z = \frac{\sin t}{\sin\phi \, \cos t - \cos\phi \, \tan\delta}, \qquad (44)$$

from which Z is to be computed after the hour-angle t has been determined.

The field operations may be conducted exactly like those explained in the last Article. Another method of observation preferred by many observers is illustrated in the notes below. Pointing is first made on the mark at B and the horizontal circle is read; then four pointings and readings are made on the star, two with the telescope in the direct position and two with it in the reverse position; finally a pointing and reading on the mark is taken again; each reading is of course the mean of the two verniers. The time as indicated by the watch must be noted for each pointing on the star, and the mean of these times is that to be used to find the hour-angle t. The process of finding t by the help of the nautical almanac is shown in the lower part of the table. Then from the formula tan$Z$ is found to be negative and hence the star was west of the meridian; accordingly Z is $-0° 59' 06''.3$, and finally the geodetic azimuth of the line AB is $175° 57' 47''.6$, the probable error of which may be estimated at $10''$ or $15''$. Making a number of observations on different parts of the circle and taking their mean, a fair determination of azimuth may be obtained by one night's work.

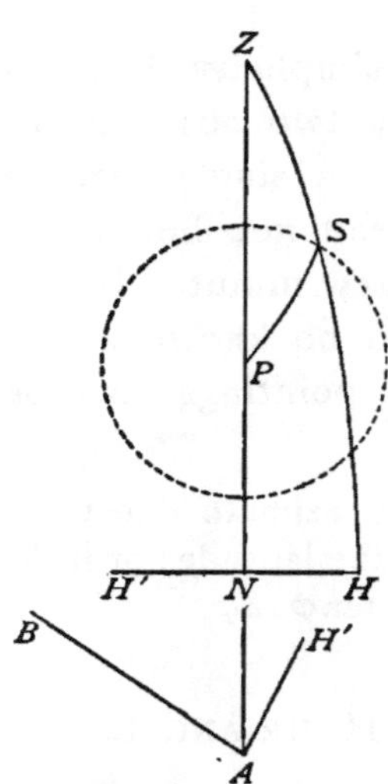

To secure the elimination of the instrumental errors of the transit more completely, one half of the pointings on the star may be made by looking at its reflection in a dish of mercury placed near the objective end of the telescope. When a geodetic theodolite is used corrections for the error of level in the telescope standards are to be applied, unless this be eliminated by taking half the observations in a mercury horizon. If a sidereal chronometer is at hand the time should be noted by it, as thus the reduction of local mean solar time to sidereal time is avoided.

AZIMUTH BY POLARIS, AUGUST 23, 1897.

Watch Time P.M.	Tel.	Readings on Mark.	Readings on Star.	Data.
	D	25° 11′ 20″		$\phi = 40° 36' 24''$
	R	25 11 20		$\lambda = 75° 22' 50''$
$8^h\ 16^m\ 30^s$	*R*		30° 14′ 30″	$\delta = 88° 45' 35''.6$
8 18 30	*D*		30 13 05	$\alpha = 1^h\ 20^m\ 19^s.2$
8 20 30	*D*		30 11 50	
8 22 30	*R*		30 10 40	Watch $5^s.5$ slower than Eastern standard time.
	R	25 11 00		
	D	25 11 10		
$8^h\ 19^m\ 30^s$		25° 11′ 12″5.	30° 12′ 31″.2	J. Doe, observer.

05.5 = Watch error.	30° 12′ 31″.2
8^h 19^m $35^s.5$ = Eastern standard time.	25 11 12 .5
− 01 31 .3 = Longitude correction.	*BAH′* = 5 01 18 .7
8 18 04 .2 = Local mean solar time.	*Z* = −0 59 06 .3
+ 01 21 .7 = Reduction to sidereal interval.	*BAN* = 4 02 12 .4
2 30 48 .5 = Sidereal time Greenwich mean noon.	180
− 00 49 .4 = Longitude correction.	Azimuth *AB* = 175° 57′ 47″.6
10^h 49^m 25^s .0 = Sidereal time.	R. Roe, computer.
1 20 19 .2 = Right Ascension of Polaris.	
9^h 29^m $05^s.8$ = Hour-angle in time.	
142° 16′ 27″.3 = *t*	

Any circumpolar star may be used by this method, but preference is generally given to Polaris as it is of second magnitude and easily identified. Other stars sometimes used are δ Ursæ Minoris and 51 Cephei, which are of fifth magnitude and hence not so easily located as Polaris.

Although theoretically the observation may be taken at any time, yet a discussion of equation (44) will show that the conditions most favorable to precision occur when t is either about 90° or 270°, that is when the star is near elongation. When the star is at either the upper or lower culmination small errors in ϕ and t may produce large errors in Z, and hence a star should not be observed when near its meridian passage. Errors due to either ϕ or δ may be eliminated by observing the star at symmetrical positions east and west of the meridian, and taking the mean of the two computed results.

Prob. 44. If an error of 15 seconds had been made in the mean time of the watch readings in the above example, what error would have been produced in the resulting azimuth of the line AB?

45. Latitude by the Sun.

When the sun is on the meridian it is at its maximum altitude very nearly, and if this be measured with a sextant the latitude of the place becomes known. Thus in the figure let the circle be a meridian section of the celestial sphere, P the pole, Z the zenith, Q the equator, S the sun, and HH the horizon. The arc SH' is the altitude of the sun, SQ is its declination, and ZQ is the latitude of the place of observation A. Let h be the meridian altitude, corrected for refraction and parallax, δ the declination of the sun, and ϕ the latitude of the place. Then, from the figure,

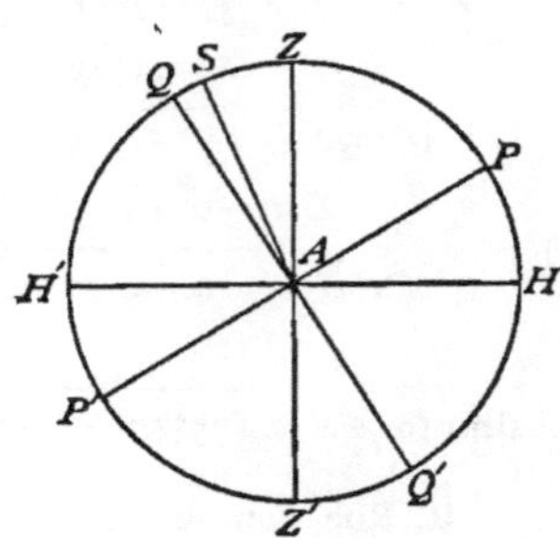

$$\phi = 90° - h + \delta, \qquad (45)$$

in which δ is positive when the sun is north of the equator and negative when it is south of the equator.

On the ocean the altitude is taken by bringing down the

image of the sun until its lower limb touches the sea horizon. On land the image is brought down until the lower limb touches its reflection as seen in a dish of mercury, and thus the double altitude is read. The operations are begun several minutes before apparent noon and a number of measurements made which give altitudes gradually increasing to a maximum and then decreasing. The proper corrections are then to be applied to the maximum altitude, and finally the above formula gives the latitude, δ being taken from the nautical almanac.

For example on October 1, 1897, the maximum double altitude of the sun's lower limb, observed with a sextant by a student at Lehigh University and corrected for index error and eccentricity of the instrument, was 91° 51′ 17″.2, one-half of which, or 45° 55′ 38″.6, is the apparent altitude of the lower limb. To this is to be added 16′ 01″.6 for the sun's semi-diameter, giving 46° 11′ 40″.2 apparent altitude of the sun's center. The refraction correction to be subtracted is 0′ 53″.1 and the parallax correction to be added is 06″.2, and thus the true altitude h was 45° 54′ 51″.6. The declination of the sun being S. 3° 28′ 44″.3, as found from the nautical almanac, taking account of the difference in longitude, the latitude of the place by (45) is 40° 36′ 24″, a result whose probable error is 5″ or more, since but a single reading was taken on the sextant.

A more precise determination can be made by taking about six altitudes at intervals of one minute, half being before and half after the time of maximum altitude. It is well also that three of them should be taken by bringing the sun's lower limb into coincidence with its image in the mercury, and three by bringing the upper limb to coincide with its image; the mean of the six then gives the double apparent altitude of the sun's center. This altitude, after correction for the errors of the instrument and for parallax and refraction, may be safely used to give a latitude determination with a probable error less than 5″, and this can be rendered smaller by combining the results of several observations made on different days. Some observers apply to each of the altitudes a correction to reduce them to the meridian, but this requires a knowledge of the local time and longitude, and the computation generally involves more labor than is justified by the precision of the vertical angles taken with a sextant or engineers' transit.

When the local mean solar time is known, or when it can be obtained from a watch indicating standard time, the observation may be made at any hour-angle with results as satisfactory as those found from noonday work. The altitudes may be taken at even minutes of time, or the sextant may be set at even minutes of angle and the watch be read to seconds of time. Half the altitudes being taken upon the lower limb and half upon the upper, the mean of all furnishes the altitude of the sun's center, and the mean of the recorded times gives the corresponding time from which the apparent solar time and the hour-angle t are found. Let V be an angle computed from

$$\tan V = \tan\delta/\cos t,$$

then the latitude ϕ is found from

$$\cos(\phi - V) = \sin V \sin h/\sin\delta. \qquad (45)'$$

For example, ten altitudes measured in about six minutes on the afternoon of September 27, 1897, at South Bethlehem, Pa., gave after correction the mean altitude $h = 36° 03' 09''.4$, and the mean of the ten times of observation was $2^h 16^m 18^s.9$ in local mean solar time. Applying the equation of time for that day, the local apparent solar time is found to be $2^h 25^m 34^s.1$, and hence the hour-angle of the sun is $t = 36° 23' 31''.5$. From the nautical almanac, knowing that the local time of the place of observation is $5^h 1^m 32^s$ slower than Greenwich time, the declination of the sun is $-1° 57'$

32''.8. Then by computation V is $-2° 25' 59''.6$ and finally the latitude ϕ is found to be $+40° 36' 24''.6$.

The probable error of a single latitude determination made in this manner is about 2'' or 3''. To obtain more precise results a star should be observed, and in general all astronomical work on the sun has a much lower degree of precision than that done on the stars.

Prob. 45. From the values of δ, t, and h, as stated above, compute the values of V and ϕ. Also using the values of δ, h, and ϕ, compute t from formula (47).

46. Latitude by a Star.

The altitude of the celestial pole above the horizon of any place is the latitude of that place. Hence when a circumpolar star crosses the meridian its altitude plus or minus its co-declination gives the latitude of the place, or

$$\phi = h \pm (90° - \delta), \qquad (46)$$

the plus sign being used for the lower culmination and the minus sign for the upper culmination. As the times of the culmination of Polaris are given in surveyors' handbooks this method is well adapted to observations upon it with the sextant or with the engineers' transit. If the altitude h_1 be observed at upper and h_2 at lower culmination, then the mean of these, each being corrected for refraction, gives the latitude, or $\phi = \frac{1}{2}(h_1 + h_2)$. Here h_1 may be itself the mean of several altitudes taken at equal intervals before and after the upper culmination and h_2 may be the mean of several similarly taken before and after the lower culmination. With a good sextant the latitude may be found by a few series of observations with a probable error of one or two seconds of angle, and with a transit to a less degree of precision.

Formula (45)' of the last Article may be used to find the latitude from an observed altitude of any star in any position, if its hour-angle t is known. When a sidereal chronometer is at hand the sidereal time of taking the altitude, diminished by the right ascension of the star, gives the hour-angle in sidereal time, and fifteen times this is the value of t in angular measure. For example on December 8, 1897, ten altitudes of Polaris were taken with a sextant in about thirteen minutes at Lehigh University, the time of each being noted on a sidereal chronometer. The mean apparent altitude, after correction for index error and eccentricity, was $41° 47' 46''.3$, and applying the refraction correction $1' 06''.3$, the mean true altitude is $h = 41° 46' 40''.0$. The mean sidereal time of the times of the ten measurements was $0^h 10^m 54^s.9$, from which is subtracted the right ascension of Polaris or $1^h 50^m 33^s.6$ to give its hour-angle $22^h 50^m 33^s.6$, whose equivalent in degree measure is $t = 342° 38' 24''.0$. The declination of Polaris being $\delta = 88° 46' 11''.9$, the auxiliary V is found to be $88° 49' 33''.5$, and then from (45)′ there results for the latitude the value $\phi = 40° 36' 17''.3$. The probable error of a single determination of latitude made in this manner is much less than that of one found from observations on the sun, say about $1''$ or $2''$. When a common watch is used its error and rate should be known so that the time corresponding to the mean altitude may be converted into local mean solar time and then, with the help of the nautical almanac, into sidereal time, from which the hour-angle t is found as before. It is preferable that the student should use a common watch rather than a sidereal chronometer, since the former is more generally at hand in actual work.

The best time for making this observation is when the star is near culmination, since then an error in h produces the smallest error in ϕ. In the above example the star was about $1^h 10^m$ from the lower culmination and hence in a favorable position.

Prob. 46. Deduce $\sin h = \sin \phi \sin \delta + \cos \phi \cos \delta \cos t$, and show that an error dh gives the least error $d\phi$ when $t = 0°$ or when $t = 180°$.

47. TIME.

A watch may be set to local apparent solar time by noting the instant when the sun attains its maximum altitude, and then, applying the equation of time, local mean solar time is approximately known. At any telegraph station in the United States a watch may be closely set to standard time, Eastern standard time being the mean solar time of the 75th meridian and hence five hours slower than Greenwich mean solar time, while Central, Mountain, and Pacific standard times are the mean solar times of the 90th, 105th, and 120th meridians respectively. The mean solar time at any other meridian is found from standard time by applying to standard time the correction for difference of longitude, 15 degrees corresponding to one hour of time. When greater precision is required an altitude of the sun or of a star is to be taken and from this the error of the watch can be computed, if the latitude of the place is known.

In the figure of Art. 42 let S be the place of the sun or star; the arc SZ is the co-altitude or zenith distance z, the arc SP is the co-declination $90° - \delta$, the arc ZP is the co-latitude $90° - \phi$, and the angle SPZ is the hour-angle of the star, which is designated by t. The solution of the spherical triangle gives

$$\cos z = \sin\phi \sin\delta + \cos\phi \cos\delta \cos t,$$

from which t can be computed; it is, however, customary to reduce this equation to the form

$$\tan\tfrac{1}{2}t = \sqrt{\frac{\sin\frac{1}{2}(z + \phi - \delta)\ \sin\frac{1}{2}(z - \phi + \delta)}{\cos\frac{1}{2}(z + \phi + \delta)\ \cos\frac{1}{2}(z - \phi - \delta)}}. \qquad (47)$$

From this t is found in degrees, minutes, and seconds, and this value is then changed into time by dividing it by 15. When the sun is observed this result is apparent solar time; when a star is observed its sidereal time interval is to be reduced to mean solar time.

For example, take the data of Art. 42 where observations on the sun gave the mean corrected co-altitude $z = 46° 02' 32''$ at $8^h 43^m 00^s$ A.M. by the watch, the sun's declination being $\delta = 19° 54' 05''$ and the latitude of the place $\phi = 40° 36' 27''$. Inserting these values in the above formula there is found $t = - 48° 32' 50''$, which is the hour-angle between the sun and the meridian; this reduced to time gives $3^h 14^m 11^s.3$ as the interval between the time of the mean altitude and that of apparent solar noon. Hence $8^h 45^m 58^s.7$ was the local apparent solar time, and subtracting the equation of time for the given day, there results $8^h 42^m 15^s.5$ as the local mean solar time which corresponded to $8^h 43^m 00^s$ of the watch. Hence the deviation of the watch from local mean solar time was $0^m 44^s.5$ fast. Further as the place of observation was $0° 22' 35''$ west of the 75th meridian, and as this corresponds to $0^h 01^m 30^s.3$, it follows that the deviation of the watch from Eastern standard time was $45^s.8$ slow. The probable error of this determination may be several seconds.

Far better work can be done by observing a star, and a good sextant is always to be preferred to an engineer's transit for taking the altitudes, the image being brought down to coincide with its reflection in a dish of mercury. The following is a record of an observation made at Lehigh University on May 9, 1899, by this method. The watch was supposed to carry Eastern standard time and it was required to determine its error. The sextant was set successively at even 10 seconds of arc and the watch time of each noted; thus the observed mean double altitude $90° 05' 00''$ occurred at $8^h 02^m 36^s.64$ by the watch. This is corrected for index error and eccentricity, and the apparent double altitude found, to which a refraction correction, computed from a formula that takes barometer and thermometer into account, is applied. Thus the true zenith distance z is found, and from this and the given latitude of the place and declination of the star the hour-angle t is computed from the above

TIME BY α GEMINORUM (CASTOR), MAY 9, 1899.

Obs. No.	Double Altitude.	Watch Time P.M.	Data and Remarks.	
1	90° 50′	$8^h\ 00^m\ 36^s.5$	J. H. O., observer. C. L. T., recorder.	
2	90 40	01 03 .5	Pistor and Martin's Prismatic Sextant with mercury horizon and glass cover.	
3	90 30	01 30 .5		
4	90 20	01 56 .5	Watch carrying approximate Eastern standard time.	
5	90 10	02 23 .5	$\phi = 40°\ 36'\ 23''.2$	Index Error, from Arcturus.
6*	90 00	02 50 .2	$\lambda = 75°\ 22'\ 57''.3$	− 16′ 45″
7	89 50	03 16 .6	$\alpha = 7^h\ 28^m\ 11^s.30$	− 16 50
8	89 40	03 42 .5	$\delta = 32°\ 06'\ 36''.7$	− 16 55
9	89 30	04 11 .4	Barometer, 29^{in} .550	− 16 55
10	89 20	04 35 .8	Attached therm., 66°.0 F.	
Means	90° 05′ 00″	$8^h\ 02^m\ 36^s.64$	Detached therm., 64.7 F.	− 16′ 51″.25

* Horizon cover reversed.

Observed $2h$ =	90° 05′ 00″
Index error =	− 16 51.25
Eccentricity =	− 48.85
$2h$ =	89 47 25.90
Apparent h =	44 53 42.95
Refraction =	− 55.46
True h =	44 52 47.49
z =	45 07 12.51
Hour-angle t =	55° 59′ 50″
Hour t =	$3^h\ 43^m\ 59^s.33$

True sidereal time $t + \alpha$ =	$11^h\ 12^m\ 10^s.63$
Sidereal time mean noon =	3 09 09 .27
Sidereal interval after mean noon =	8 03 01 .36
Correction to mean solar time =	− 01 19 .29
Local mean solar time =	8 01 42 .07
Reduction to 75th meridian =	+ 01 31 .82
Eastern standard time =	8 03 13 .89
Watch time =	8 02 36 .64
Watch error (slow) =	$37^s.25$

C. L. T., computor.

formula, and converted into time measure. The true sidereal time then results by adding the right ascension of the star, and this is converted into local mean solar time and then into Eastern standard time, from which finally the watch error is found to have been 37.25 seconds slow. The probable error of this determination is less than one-quarter of a second.

The most favorable position of a star for this work is when

it is on the prime vertical. For, if dz be an error in z, the corresponding error dt in t, obtained by differentiating the first equation of this Article, is $\sin z \cdot dz/\cos\phi \cos\delta \sin t$, and since $\sin z/\cos\delta \sin t = \sin Z$, this gives $dt = dz/\cos\phi \sin Z$. Accordingly the azimuth Z should be 90° or 270° in order that dt may have its smallest value. In the same manner it is shown that errors in the assumed latitude produce the least effect when the star is on the prime vertical.

Prob. 47. Using the data of the above observation on Castor, find the error in the computed Eastern standard time which would be caused by an error of 01″ in the altitude h; also the error which would be caused by an error of 01″ in the latitude ϕ.

48. Longitude.

When accurate standard time is at hand the comparison of it with the local mean time gives the longitude. Thus, if the local mean solar time of a place has been found by a star observation to be $14^{m}\ 08^{s}.4$ faster than Central standard time, the place is 3° 22′ 04″ east of the 90° meridian and hence its longitude is 86° 27′ 56″ west of Greenwich. This method is used at sea, where daily observations for local mean solar time are made on the sun or stars when the weather permits, this local time being compared with a chronometer which indicates either Greenwich mean solar time or that of a port whose longitude is known. As one second of time is equivalent to fifteen seconds of angle, it is seen that this method is not very precise, particularly when it is considered that the best watches are liable to vary one or more seconds per day.

The method of lunar distances is extensively used at sea for finding the Greenwich time. In the nautical almanac will be found the true angular distances between the moon's center and several stars and planets for every day in the year and for three-hour intervals, these distances being stated as they would appear from the center of the earth. If one of these apparent distances be measured at any place, as also the apparent altitudes of the star and moon, the data are at hand for computing the true distance as seen from the center of the earth at the same instant, and thus from the almanac Greenwich mean time is known. Then, the difference between local and Greenwich time gives the longitude of the place. This method involves laborious computations unless special tables are at hand.

Another method is that of lunar culminations which requires that azimuth and time should have been determined. The instant of the passage of the moon's bright limb across the meridian is observed, and a correction applied to find the local mean time of passage of the moon's center. This local mean time, converted into sidereal time, furnishes the right ascension of the moon, while the Greenwich mean time corresponding to the same right ascension can be found from the almanac. Lastly the difference between local and Greenwich mean time gives the longitude of the place.

As an example of the method of lunar culminations the following rough observation with an engineer's transit, made at Lehigh University on May 23, 1899, may be of interest. On that day the moon crossed the meridian at about $10^h\ 55^m$ P.M., and it was accordingly arranged to determine azimuth by pointing on Polaris a few minutes previous. By a simple computation it was determined that the azimuth of Polaris at $10^h\ 45^m$ local mean time was $180°\ 36'\ 23''.4$ and at that instant the cross-hair of the transit telescope was set on the star. Then the angle $0°\ 36'\ 20''$ was turned off toward the west and the telescope reversed, thus pointing southward in the plane of the meridian. When the moon's west limb touched the vertical cross-hair the time was noted as $10^h\ 53^m\ 34^s.1$. Reducing this to sidereal time with the assumed longitude 5^h, and adding a correction for the time required for the semi-diameter of the moon to pass the meridian, the right ascension of the moon's center when crossing the meridian of the place is found to be $15^h\ 00^m\ 53^s.44$, and the corresponding mean local time $10^h\ 54^m\ 55^s.59$. From the nautical almanac the Greenwich mean time at which the moon's center had this right ascension is found to be $15^h\ 54^m\ 27^s.84$, and consequently the longitude of the place of observation is $4^h\ 59^m\ 42^s.25$ in time or $74°\ 55'\ 34''$ in arc, a result which is in error by nearly half a degree, the true value being $75°\ 22'\ 23''$.

It thus appears that no close determination of the longitude of a place can be made by the method of moon culminations with an engineer's transit. Nevertheless in an unexplored region the method is of value in making an approximate determination to be used in time observations and in taking quantities from the nautical almanac.

Prob. 48. At a certain place on December 5, 1900, the moon's right ascension was observed as $4^h\ 32^m\ 05^s.31$ at 8 P.M. local mean time. From the nautical almanac it is found that the right ascensions $4^h\ 31^m\ 03^s.69$ and $4^h\ 33^m\ 37^s.82$ occurred at 3 A.M. and 4 A.M., Greenwich mean time, on December 6, 1900. Find the longitude of the place of observation.

49. Precise Determinations.

The methods set forth in the preceding pages give results whose precision is far lower than that needed for the astronomical work of a geodetic survey. When it is required to determine azimuth, latitude, and longitude at one of the stations of a geodetic triangulation such methods are generally used to furnish preliminary approximate values, for it has been seen that each of these elements depends upon the others, and hence rough methods must precede precise ones. These preliminary values may be supposed to give the latitude within one or two seconds, the longitude within ten or twenty seconds, and the azimuth within six or eight seconds.

To make a precise determination of azimuth a direction theodolite, having a circle divided to 5 minutes and reading by microscopes to seconds or less, is used. The observations are made on close circumpolar stars by the method of Art. 35, great pains being taken to eliminate the error of level in the horizontal axis of the telescope. By making a sufficient number of measurements the azimuth of a line running from the station to a signal may be found with a probable error of 1″ or less, and by measuring the angle between this line and one of the sides of the triangulation the azimuth of the latter is known with almost equal precision.

To make a precise determination of latitude a zenith telescope is to be set up in the plane of the meridian and the difference of the meridian zenith distances of two stars that cross the meridian near the zenith, but on opposite sides of it, is observed. Let δ_1 and δ_2 be the declinations of the two stars, the first being south of the zenith, z_1 and z_2 their apparent zenith distances, and r_1 and r_2 their refraction corrections. Then for the first star $\phi = \delta_1 + z_1 + r_1$ and for the other $\phi = \delta_2 - z_2 - r_2$. The addition of these gives

$$2\phi = \delta_1 + \delta_2 + (z_1 - z_2) + (r_1 - r_2),$$

so that it is only necessary to measure $z_1 - z_2$ by the micrometer in the field of the zenith telescope and then apply the small refraction correction. By this method it is easy to determine latitude with a probable error less than 0″.1.

To make a precise determination of longitude a telegraph line must connect the station with an observatory whose longitude is known. A portable astronomical transit instrument is mounted in the plane of the meridian and the time of passage of several equatorial stars is signalled by the telegraph line to the observatory. When the same stars pass the meridian of the observatory their time of passage is signalled to the station. A single clock in the telegraph circuit may be used to make a chronographic record of both series of signals, and thus the difference in time is known, from which the longitude directly results. The probable error of the difference of longitude thus determined may be made as small as $0^s.01$ or 0″.15.

Numerous observations made at many different observatories have established the fact that small periodic changes in the latitudes of all places are constantly going on. This is due to a slight wabbling motion of the earth's mass with respect to its axis, so that the axis performs an apparent revolution around its mean position in about 425 days, and consequently the north pole of the earth makes a similar revolution around its mean position. The radius of this circle varies from $0''.16$ to $0''.36$, and consequently the latitude of any given point on the earth's surface may vary from $0''.32$ to $0''.72$ at different times. It is hence seen that decimals of seconds occurring in common latitude determinations have no definite meaning. As all methods for determining the azimuth of a line involve a knowledge of the latitude of that end where the observation is made, it follows that the astronomical azimuths of all lines on the earth's surface also undergo periodic changes; and the same holds true for longitudes of places. These changes will be the greater the nearer the line or place is to the north pole, but near the equator they will be very small.

In the science of geodesy the words azimuth, latitude, and longitude have a signification slightly different from that in astronomy, as will be seen in Chapter VII. These geodetic elements enable a fixed system of coordinates to be established by which the relative positions of points on the earth's surface can be expressed to a degree of precision limited only by our knowledge of the shape and size of the earth.

Prob. 49. Consult Albrecht's Bericht über den Stand der Erforschung der Breitenvariation (Berlin, 1899), and give a sketch showing how the true north pole moves around its mean position.

CHAPTER VI.

SPHERICAL GEODESY.

50. EARLY HISTORY.

Geodesy is the science that sets forth the principles and methods whereby large areas on the surface of the earth may be surveyed and mapped with precision. If the surface of the earth were a plane, as certain ancient peoples supposed, the science of geodesy could never have arisen, since the elementary geometry of Euclid would be capable of measuring and representing its geographical features. In fact, however, measurements conducted upon this supposition become more or less entangled in discrepancies according to the size of the country over which they are carried. For instance, let three points be taken on the earth's surface at considerable distances apart; the sum of the three angles of the triangle thus formed is found, if measured by an instrument whose graduated arc is placed level at each station, to be greater than 180 degrees. From these and many other discrepancies it is to be concluded that the earth's surface is not a plane.

Many facts are known from which it is inferred that the earth is globular, such as the appearance of the top of a light-house earlier than its base to a ship approaching the shore, the dip of the sea horizon, the elevation of the pole star as we travel north and its depression as we travel south, the analogy of the other planets which seem to be globular when viewed through a glass; and the circular form of the earth's shadow as observed in a lunar eclipse. To these must be added the well-known circumstance that travellers, going ever eastward, pass entirely around the earth and return to the point of starting. From these facts it is concluded that the earth is globular, that is to say like a globe, but whether spherical, spheroidal, or ellipsoidal, there is thus far no evidence.

The surface whose size and shape is to be investigated in the following pages is that of the great ocean which covers fully three-fourths of the globe. Although this is agitated by winds and raised in tides its mean level can be accurately determined. Moreover the land is really elevated but little above the ocean, for it is now known that the radius of the earth, regarded as a sphere, is nearly 4 000 miles, while the highest mountains rise only about 5 miles. Hence measurements made upon the land can at the utmost cause an error of only one eight-hundredth part in the value of the radius.

The early Greek philosophers speculated upon the shape of the earth. Anaximander (570 B.C.) called it a cylinder whose height was three times its diameter, the land and sea being on its upper base. Plato (400 B.C.) thought it a cube. Aristotle (340 B.C.) gives reasons for supposing it to be a sphere and mentions, as also does Archimedes (250 B.C.), that geometers had estimated its circumference at 300 000 stadia. The first recorded observations for determining the size of the sphere are, however, those made in Egypt by Eratosthenes (230 B.C.); his method, though rude in measurement, is correct in principle, and from it he concluded that the circumference of the earth was 250 000 stadia.

The process by which Eratosthenes deduced the size of the earth will now be described. He noticed that at Syene in southern Egypt the sun on the day of the summer solstice cast no shadow of a vertical object, while at Alexandria in northern Egypt the rays of the sun on the same day of the year made an angle with the vertical of one-fiftieth of four right angles. From this he concluded that the distance between Syene and Alexandria was one-fiftieth of the circumference of the earth, and as that distance was about 5 000 stadia he claimed the whole circumference to be 250 000 stadia. The exact length of the stadium is now unknown, so that the precision of his result cannot be judged, yet the name of Eratosthenes will ever be honored in science as the originator of the method of deducing the size of the earth from a measured meridian arc.

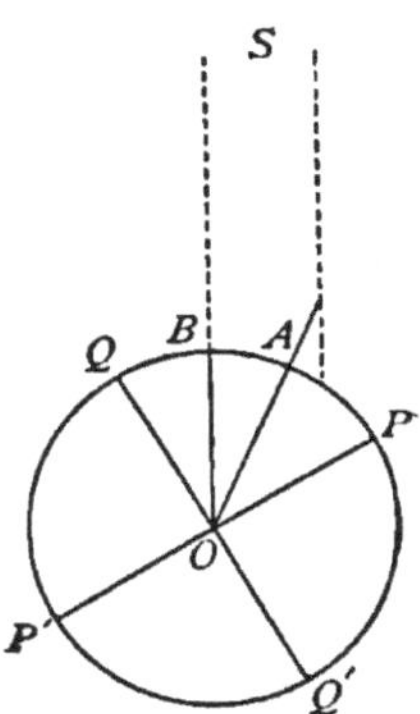

To explain the reasoning of Eratosthenes let the figure represent a meridian section of the earth, PP' being the axis, QQ' the equator, P the north pole, A the position of Alexandria, and B that of Syene, while AS and BS give the directions of the sun at the summer solstice. Assuming that the section is a circle and that the rays of the sun are parallel it is clear that the angle BOA is equal to the angle which the sun's rays make with OA. Thus, if this angle be $\frac{1}{50}$th of 360 degrees, it follows that the circumference is fifty times the distance AB. The reasoning of Eratosthenes hence involves two fundamental conceptions besides those of geometry, namely, that the earth is a sphere and that the sun is at a great distance from it.

The method of Eratosthenes is called the measurement of a meridian arc. Thus, let the distance between the two points A and B be l, and let the angle AOB be θ degrees; then in modern reasoning,

$$l/\theta = \text{length of one degree of the meridian,}$$
$$360l/\theta = \text{length of circumferences of the earth,}$$
$$57.2958l/\theta = \text{length of radius of the earth.}$$

Eratosthenes found the distance l from the statements of

travellers, later observers rolled a wheel, or measured it with a chain, but the modern method is to compute it from a precise triangulation. Eratosthenes found the angle θ from the shadows cast by vertical posts, but later observers found it from the latitudes of the places; for the angle QOA is the latitude of A, while QOB is the latitude of B, and hence θ is the difference in latitude between the two ends of the meridian arc.

For several hundred years after the time of Eratosthenes the doctrine of the spherical form of the earth was generally accepted by astronomers. Posidonias (90 B.C.) measured the meridian arc between Alexandria and Rhodes, using a star to determine the latitudes and deduced 240 000 stadia for the circumference. But this knowledge of the Greeks was all lost as their civilization declined, and for more than a thousand years Europe, sunk in intellectual darkness, made no inquiry concerning the size or shape of the earth. Only in Arabia were the sciences at all cultivated during this period. There the Caliph Almamoun summoned astronomers to Bagdad, and one of their labors was the measurement, on the plains of Mesopotamia, of an arc of a meridian by wooden rods, from which they deduced the length of a degree to be 56⅔ Arabian miles, or probably about 71 English miles.

Prob. 50. If the earth is represented by a sphere 16 inches in diameter, what is the height in inches of the tallest mountain?

51. History from 1300 to 1750.

In the year 1322 a traveller named Mandeville published a volume describing his journeys; this is generally regarded as the earliest English prose work. In it is a lengthy and labored argument to prove that the "lond and the see ben of rownde schapp and forme" and that the circumference of the earth has 360 degrees like that of the heavens. He concludes, "be the Earthe devysed in als many parties, as the Firmament; and lat every partye answere to a Degree of the Firmament; and wytethe it wel, that aftre the auctoures of Astronomye, 700 Furlonges of Earthe answeren to a Degree of the Firmament; and tho ben 87 Myles and 4 Furlonges. Now be that here multiplyed by 360 sithes; and than thei ben 31 500 Myles, every of 8 Furlonges, aftre Myles of oure Contree. So moche hathe the Earthe in roundnesse, and of heighte environ, aftre myn opynyoun and myn undirstondynge."

These views of Mandeville appear to have produced but little influence, for it was not until the fifteenth century, when the first gleams of light broke in upon the darkness of the middle ages, that men began to think again about the shape and size of the earth. Navigators began to doubt that its surface was a level plane, and here and there one, like Columbus, asserted it to be globular. In the sixteenth century, the doctrine of the spherical form of the earth was again generally accepted, and one of the ships of Magellan, after a three years' voyage, accomplished its circumnavigation. With the acceptance of this idea arose also the question as to the size of the globe, and Fernel, in 1525, made a measurement of an arc of a meridian by rolling a wheel from Paris to Amiens to find the distance and by observing the latitudes with large wooden triangles, from which he deduced about 57 050 toises for the length of one degree. At this time methods of precision in surveying were entirely unknown. In 1617 Snellius conceived the idea of triangulating from a known base line, and thus, near Leyden, he measured a meridian arc which gives 55 020 toises for the length of a degree. Norwood, in 1633, chained the distance from London to York, and deduced 57 424 toises for a degree. Picard, who was the first to use spider lines in a telescope, remeasured, in 1669, the arc from Paris to Amiens, using a base line and triangulation, and found one degree to be

57 060 toises. This was the result that Newton used when making his famous calculation which proved that the moon gravitated toward the earth.

The toise, it should here be noted, was an old French measure, approximately equal to 6.3946 English feet or 1.949 meters. It is of classic interest on account of its use in all the early meridian arcs and in the surveys for deciding upon the length of the meter.

From 1690 to 1718 Cassini carried on surveys in France, more precise probably than any preceding ones, and in 1720 he published the following results regarding three meridian arcs:

Arc.	Mean Latitude.	Toises in One Degree.
1	49° 56′	56 970
2	49 22	57 060
3	47 55	57 098

From these it appeared that the length of a degree of latitude increased toward the equator, or that the earth was flatter at the equator than at the poles. In other words he claimed that the earth was not spherical but spheroidal, and that the spheroid was a prolate one. From the time men had ceased to believe in the flatness of the earth, and had begun to regard it as a sphere, their investigations had been directed toward its size alone; now, however, the inquiry assumed a new phase, and its shape came up again for discussion.

A prolate spheroid is generated by an ellipse revolving about its major axis, and an oblate spheroid by an ellipse revolving about its minor axis. The first diagram of the figure represents a meridian section of the earth regarded as a prolate spheroid, and the second shows the section of an oblate spheroid. In each diagram PP is the axis, QQ the equator, and A a place of observation whose horizon is AH, zenith Z, latitude ABQ, and radius of curvature AR. Now if the earth be regarded as a sphere and its radius be found from a meridian arc near A, the value AR will result. In the prolate spheroid the radius of curvature is least at the poles and greatest at the equator, and the reverse in the oblate. Hence if the lengths of the degrees of latitude decrease from the equator to the poles, it shows that the

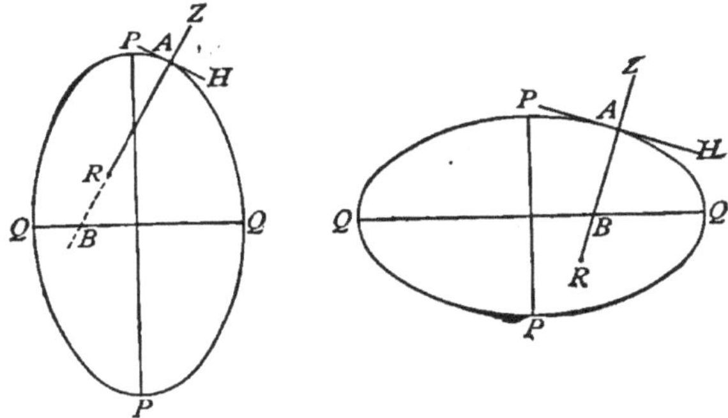

earth is prolate; but if they increase from the equator toward the poles, it is a proof that it is oblate in shape.

It is now necessary to go back to the year 1687, the date of the publication of the first edition of Newton's Principia. In Book III of that great work are discussed the observations of Richer, who, having been sent to Cayenne, in equatorial South America, on an astronomical expedition, noted that his clock, which kept accurate time in Paris, there continually lost two seconds daily, and could only be corrected by shortening the pendulum. Now, the time of oscillation of a pendulum of constant length depends upon the intensity of the force of gravity, and Newton showed, after making due allowance for the effect of centrifugal force, that the force of gravity at Cayenne, compared with that at Paris, was too small for the hypothesis of a spherical globe; in short, that Cayenne was further from the center of the globe than Paris, or that the earth was an oblate spheroid flattened at the poles. He computed, too, that the amount of this flattening at both poles was between $\frac{1}{180}$ and $\frac{1}{300}$ of the whole diameter. Now it will be remembered that Newton's philosophy did not gain ready acceptance in France; this investigation, in particular, called forth much argument, and when Cassini's surveys were completed, indicating a prolate spheroid, the discussion became a controversy. Then the French Academy resolved to send expeditions to measure two meridian arcs that would definitely settle the matter, one near the equator and another as far north as possible.

Accordingly two parties set out in 1735, one for Lapland, the other for Peru. The Lapland expedition measured its base upon the frozen surface of a river, executed its triangulation and latitude observations, and returned in two years with the results $l = 92\,778$ toises, $\theta = 1^\circ.6221$. The Peruvian expedition measured two bases, executed its triangulation and latitude work, and returned in seven years with the results $l = 176\,875$ toises, $\theta = 3^\circ.1176$. From these the values of the length of one degree were found, and then the following results could be written:

Arc.	Mean Latitude.	Toises in One Degree.
Lapland	N. 66° 20′	57 438
France	N. 49 22	57 060
Peru	S. 1 34	56 728

These figures decided the question. Since that time every one has granted that the earth is an oblate spheroid rather than a sphere or an prolate spheroid.

Prob. 51. From the above data compute the radius of curvature for the Lapland arc and for the Peruvian arc.

52. Measurement of Meridian Arcs.

The general principles regarding the measurement of a meridian arc have been given in Art. 50, but it is now to be noted that the successful execution of the work demands accurate instruments, good observers, and long-continued labor. The latitude observations are now made by the zenith telescope method of Art. 49, the bases, angles, and azimuths are measured with corresponding precision, while the adjustment by the Method of Least Squares reduces the residual errors to a minimum. In the last century these precise methods were unknown, yet the results deduced gave valuable information and progress was constantly made in methods of observation and computation. It will be of historic interest, perhaps, to give a brief account of the first meridian arc measured in the United States.

In 1763 the proprietors of Pennsylvania and Maryland employed two astronomers named Mason and Dixon to locate the boundary lines between their respective possessions. This occupied several years, and while engaged upon it, Mason and Dixon noted that several of the lines, particularly the one between Maryland and Delaware, were well adapted to the determination of the length of a degree, being on low and level land, and deviating but little from the meridian. Representing this to the Royal Society of London, of which they were members, they received tools and money to carry on the work. The measured lines are shown in the annexed sketch. AB is the boundary between Delaware and Maryland, about 82 miles long and making an angle of about four degrees with the meridian; BD is a short line running nearly east and west; CD and PN are meridians about five and fifteen miles in length respectively; CP is an arc of the parallel, the same in fact as that of the southern boundary of Pennsylvania. In 1766 Mason and Dixon set up a portable astronomical instrument at A, the southwest corner of the present State of Delaware, and by observing equal altitudes of certain stars, determined the local time and the meridian, after which the azimuth of the line AB was measured, and the latitude of A found by observing the zenith distances of several stars as they crossed

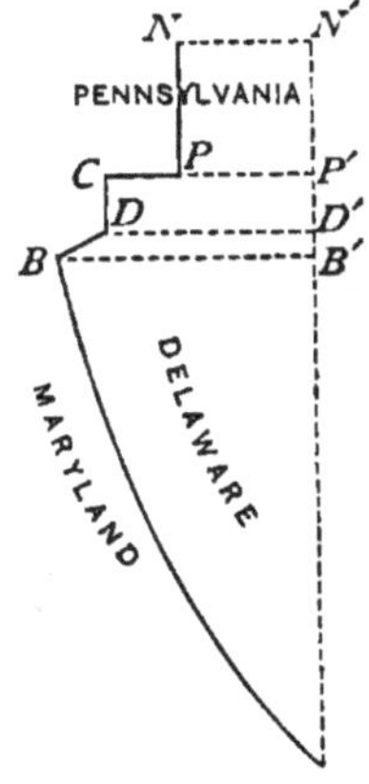

the meridian. At N, a point in the forks of the river Brandywine, the zenith distances of the same stars were also measured, from which it was easy to find the latitude of N, and the difference of latitude between A and N. In 1768 they made the linear measurements by means of wooden rectangular frames 20 feet in length. All the lines had in previous years been run in the operation for establishing the boundaries, and along each of them "a vista" cut, which "was about eight or nine yards wide, and, in general, seen about two miles, beautifully terminating to the eye in a point." Toward this point they sighted the wood frames, made them truly level and noted the thermometer in order to correct for the effect of temperature. Through the swamps they waded with the wooden frames, but across the rivers they found the distance by a measured base and triangle.

The results of this field work, as sent to England in 1768, were as follows: latitude of $A = 38° \ 27' \ 34''$, latitude of $N = 39° \ 56' \ 19''$, azimuth of AB at $A = 176° \ 16' \ 30''$, angle $CDB = 93° \ 27' \ 30''$, $AB = 434\,011.6$ feet, $BD = 1\,489.9$ feet, $DC = 26\,608.0$ feet, $PN = 78\,290.7$ feet, DC and PN being true meridians while CP was an arc of the parallel.

From these results the difference of latitude between A and N is $\theta = 1°.47917$. To find the linear distance l, an approximate value of the radius of the earth was assumed and each of the measured lines projected upon the meridian AN' by arcs of parallels NN', PP', etc. Thus were found $AB' = 433\,078.8$ feet, $B'D' = 89.8$ feet, $D'P' = 26\,608.0$ feet, and $P'N' = 78\,290.7$ feet, whose sum is $l = 538\,067.3$ feet. The length of one degree of the meridian now is

$$l/\theta = 363\,764 \text{ feet} = 68.894 \text{ miles},$$

from which the radius of curvature is

$$R_1 = 57.2958 l/\theta = 3947.4 \text{ miles}.$$

These are the final results of the measurement of the

meridian arc made by Mason and Dixon; they are now known to be too small, the present accepted values for the mean latitude of the arc being 68.984 miles and 3 952.4 miles, but in view of the primitive methods employed it is surprising that the agreement is so close.

During the fifty years following 1750 a number of meridian arcs were measured, one in South Africa, one in Italy, one in Hungary, one in Lapland, while in France and England geodetic surveys furnished the data for computing other arcs. Most important of all was the triangulation executed in France and Spain about 1800 for determining the length of the meter, which embraced an arc of ten degrees in length. All these arcs confirmed the conclusion that the earth is not a sphere, but an oblate spheroid flattened at the poles.

Prob. 52. Compute the length of a quadrant of the meridian in meters, using the results of Mason and Dixon and supposing the earth to be a sphere.

53. THE EARTH AS A SPHERE.

Although the earth is not a sphere it is sufficient in many investigations to regard it as such, since the amount of flattening at the poles is not large. In fact, if the earth is represented by a globe sixteen inches in equatorial diameter the polar diameter would be 15.945 inches, so that the difference between the two diameters would not be perceptible to the eye. The question now arises as to what value shall be taken for the radius of the earth and what is the mean length of a degree of latitude on its surface. This question cannot be answered without anticipating to a certain extent some of the conclusions of the next chapter.

The mean length of a degree of latitude is the mean of the lengths of all the degrees from the equator to the poles, or one-ninetieth of the elliptical quadrant. The value adopted for the quadrant in this book is

$$q = 10\,001\,997 \text{ meters} = 32\,814\,886 \text{ feet},$$

and from this is deduced the following useful table of mean lengths of arcs on the earth's meridian:

One degree =	111 133	meters =	364 610 feet,
One minute =	1 852.2	meters =	6 076.8 feet,
One second =	30.87	meters =	101.28 feet.

The mean length of one degree may also be stated in round numbers, easy to remember, as 69 statute miles or 60 nautical miles, one nautical mile thus being one minute of latitude.

The mean radius of the earth, considered as a sphere, must be the arithmetical mean of all the radii of the spheroid. This is evidently the same as the radius of a sphere having a volume equal to the volume of the spheroid. Let a be the equatorial and b the polar radius of the oblate spheroid, whose accepted values are 6 378 278 and 6 356 654 meters respectively; its volume is $\frac{4}{3}\pi a^2 b$. Let R be the radius of the sphere whose volume is $\frac{4}{3}\pi R^3$. Equating these values, there is found

$$R = 6\,371\,062 \text{ meters} = 20\,902\,416 \text{ feet},$$

or, in round numbers,

$$R = 6\,371 \text{ kilometers} = 3\,959 \text{ statute miles},$$

for the mean radius of the earth.

This mean value of the radius is, however, incongruous with the above mean length of a degree of latitude, for the quadrant of a circle corresponding to a radius of 6 371 kilometers is nearly six kilometers greater than the true elliptical quadrant. In certain cases it might be more logical to use the radius of a circle whose quadrant is equal to the true quadrant; this requires the equation $\frac{1}{2}\pi R = 10\,001\,997$ meters, from which

$$R = 6\,369 \text{ kilometers} = 3\,957 \text{ statute miles},$$

and this is less by two miles than the mean radius of the sphere. This discrepancy is unavoidable, since the properties of a sphere and a spheroid are not the same. Thus it is impossible, when precision is demanded, to regard the earth as a sphere.

Prob. 53. Taking the area of the earth's spheroidal surface as 196 940 400 square miles, find the radius of a sphere having the same area.

54. Lines on a Sphere.

The intersection of a plane and a sphere is always a circle. When the plane passes through the center of the sphere the circle is called a great circle, its radius being R and its circumference $2\pi R$. When the plane does not pass through the center the radius of the circle is less than R, say r, and its circumference is $2\pi r$. All great circles cut out by planes passing through the axis of the earth are called meridians and these, of course, converge and meet at the poles. All small circles cut out by planes perpendicular to the axis are called parallels. Latitude is measured north and south on the meridians from the equator toward the poles, while longitude is measured east and west on the parallels from the meridian of Greenwich.

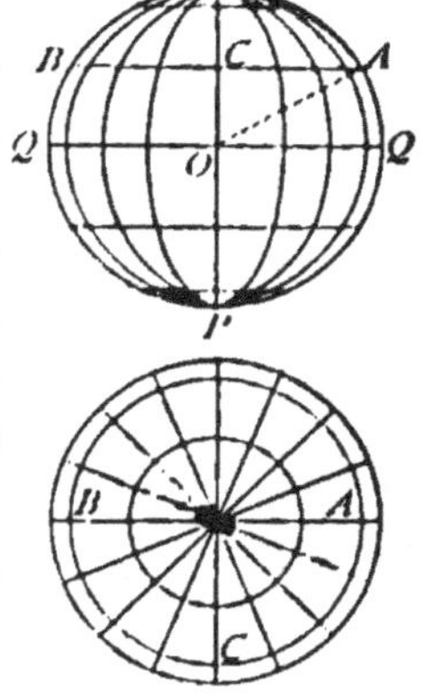

Using the mean figures of the last Article, one minute of latitude corresponds to 1 852 meters or 6 077 feet. One minute of longitude on the equator has the same value, but one minute of longitude on any parallel circle is smaller the nearer the circle is to the pole. Thus if A be a point on a parallel whose radius AC is r, and whose latitude AOQ is ϕ, and if R be the radius of the sphere, then $r = R\cos\phi$, and accordingly $2\pi r = 2\pi R\cdot\cos\phi$, that is, the length of the parallel circle is equal to the length of a great circle multiplied by the cosine of the latitude. Hence the length of one degree or one minute of longitude at any latitude is found by multiplying the values of the last Article by the cosine of the latitude. Thus, using 1 852 meters or 6 077 feet for the length of one minute at the equator, the length of one minute of longitude at latitude 40° is 1 419 meters or 4 655 feet, while at latitude 80° it is 322 meters or 1 055 feet.

The above figure shows two orthographic projections of the meridians and parallels of a sphere, the first being a projection on a plane through the axis, and the other a projection on the plane of the equator. The parallels appear as straight lines in the first diagram and as circles in the second.

The shortest distance on the surface of a sphere between any two points is along an arc of a great circle joining them. This can be rigidly demonstrated by establishing a general expression for the length of a line on the spherical surface and making it a minimum, but it will be just as well for the student to satisfy himself of the truth of the proposition by actually drawing and measuring lines on such a surface. As an illustration, the distance from A to B in the above figure may be computed by the route ACB along the parallel and by the route APB on the great circle. The length of the first is $\pi R \cos\phi$ and that of the second is $\pi R - 2\phi R$. Thus, if ϕ be 45 degrees or $\frac{1}{4}\pi$ radians, the first route has the length $2.22R$ while the second has the length $1.57R$. In like manner the distance from A to C is $1.11R$ along the parallel, but $1.00R$ along a great circle passing through A and C.

The azimuth of a line on a sphere is estimated, as in a plane, from the south around through the west; thus the northward azimuth of all meridians is 180 degrees. As all meridians converge at the poles the back azimuth of an oblique line is not equal to its front azimuth plus 180°. A great circle passing through C with an east and west direction at that point cuts the neighboring meridians at different angles and finally crosses the equator and attains the same southern latitude as C on the opposite side of the sphere. All the meridians cut the equator at right angles, but they cut other parallels at smaller angles. An oblique line crossing all meridians at the same angle is of a spiral nature and is called a loxidrome.

Prob. 54. What part of the surface of a sphere is north of north latitude 60 degrees?

55. ANGLES, TRIANGLES, AND AREAS.

A spherical angle is the plane angle between the tangents to the arcs of the great circles at their point of intersection; thus the spherical angle BAC is the same as the plane angle bAc. When a horizontal angle is measured at a station A on the surface of the earth, the limb of the instrument is made level or tangent to the spherical surface, and hence when pointing is made upon B and C the plane angle bAc is the result of the work. If the triangle be of sufficient size it will be found that the sum of the three measured angles is greater than 180 degrees.

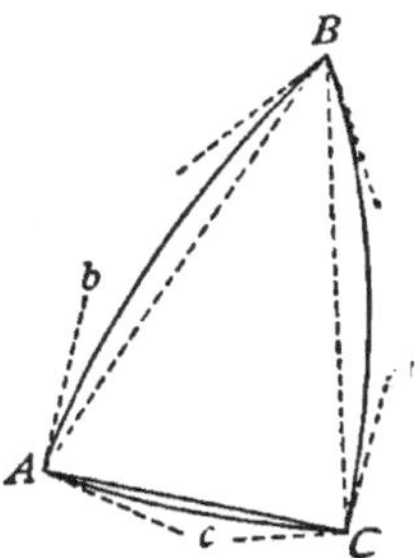

A spherical triangle is one included by three arcs of great circles. It is a well-known geometrical theorem that the sum of the angles of a spherical triangle is greater than two right angles, and that the excess above two right angles bears the same ratio to a right angle as the area of the triangle bears to the area of the tri-rectangular triangle. The tri-rectangular triangle, shown by PQO in the figure of the last Article, is one-eighth of the surface of the sphere or $\frac{1}{2}\pi R^2$. Thus from the theorem the spherical excess is given by

$$\text{Excess in right angles} = \text{area of triangle}/\tfrac{1}{2}\pi R^2,$$

or, since there are $90 \times 60 \times 60$ seconds in a right angle,

$$\text{Excess in seconds} = 648\,000 \text{ area}/\pi R^2. \qquad (55)$$

Taking for R the mean value of the radius of the earth considered as a sphere (Art. 53), this becomes

$$\begin{aligned}\text{Excess in seconds} &= \text{area in square kilometers}/197 \\ &= \text{area in square miles}/76,\end{aligned} \qquad (55)'$$

which are convenient approximate rules for practical use. Thus a triangle has one second of spherical excess for each 197 square kilometers or 76 square miles of area.

The same rule applies to quadrilaterals or polygons on the earth's surface bounded by great circles, the word excess meaning the excess of the sum of the interior spherical angles over the theoretic sum for a plane figure. Thus a polygon or triangle of the size of the State of Connecticut has a spherical excess of about 64 seconds; this amount is rarely exceeded in the triangles of geodetic triangulations and is usually much smaller.

A geodetic triangle is necessarily small since the stations must be intervisible, and hence its curved surface does not sensibly differ in area from that of the plane triangle formed by the chords of the spherical arcs. These chords are the distances computed from the triangulation work, and the corresponding plane angles are found by subtracting one-third of the spherical excess from each spherical angle. For instance, let two sides of a triangle be $a = 36\,440$ meters, $b = 23\,700$ meters, and their included angle $C = 49° \; 05'$; then the area is $\frac{1}{2}ab \sin C = 326.3$ square kilometers, and by (55)′ the spherical excess is $01''.66$.

It will be seen later that the above equation (55) is directly applicable to triangles on a spheroid by taking for R the radius of the sphere osculatory to the spheroid at the center of gravity of the triangle. In many common cases, however, the rough rules of (55)′ will give the spherical excess correctly to hundredths of a second.

The area of a zone of the sphere bounded by the parallel circles whose latitudes are L_1 and L_2 is easily derived. The differential expression is $2\pi rRdL$, where r is the radius of the parallel and R that of the sphere. But $r = R\cos L$, and hence

$$A = 2\pi R^2 \int \cos L\, dL = 2\pi R^2 (\sin L_2 - \sin L_1)$$

is the area of the zone between the upper latitude L_2 and the lower latitude L_1. Thus to find the area between latitude 30° and the equator, $L_2 = 30°$ and $L_1 = 0°$, whence $A = \pi R^2$ or one-sixteenth of the surface of the sphere.

The area of a trapezoidal degree, that is, of a surface bounded by two parallels one degree apart and by two meridians one degree apart, may be readily deduced from the last equation and will be found $0.0003046 14 R^2 \sin L$, in which L is the middle latitude of the trapezoid. Thus, taking $R =$ 3 959 miles and $L = 45°$, the area of the trapezoidal degree is 3 376 square miles.

Prob. 55. In a spherical triangle two angles are observed to be 79° 03′ 41″.93 and 59° 35′ 44″.38, and the included side is 23 700 meters. Compute the spherical excess and find the other spherical angle.

56. Latitudes, Longitudes, and Azimuths.

Let A and B be two points on the surface of the sphere, L and M being the latitude and longitude of A, and L' and M' those of B. The latitudes are estimated northward from the equator and the longitudes westward from the meridian of Greenwich, both in degrees, minutes, and seconds of arc. Let a great circle connect the points A and B, and let its angular length be S. Let the meridians through A and B be produced to meet at the north pole P and to cross the equator at Q and Q.′

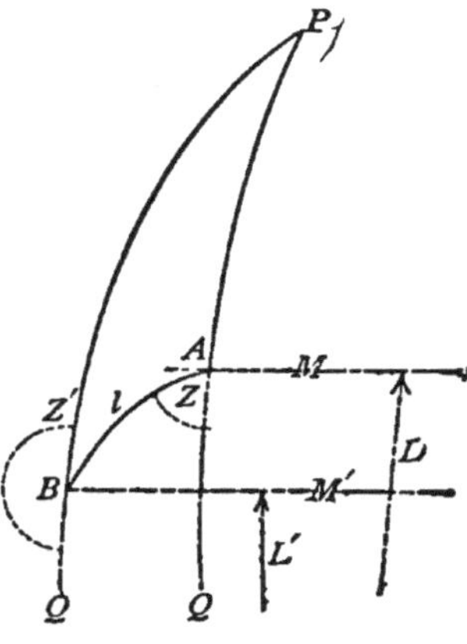

The azimuth of AB is the angle QAB, which is called Z, and the azimuth of BA is the obtuse angle $Q'BA$, which is called Z'. Let the latitude and longitude of A be given, together with the length and azimuth of AB. It is required to find the latitude and longitude of B and the azimuth of BA.

In the spherical triangle ABP the side PA is $90° - L$, the side PB is $90° - L'$, and the side AB is S; the angle A is $180° - Z$, the angle B is $Z' - 180°$, and the angle P is $M' - M$. Writing the formula of spherical trigonometry for the cosine of PB in the notation here used, it becomes

$$\cos(90°-L') = \cos(90°-L)\cos S + \sin(90°-L)\sin S\cos(180°-Z),$$

which reduces immediately to

$$\sin L' = \sin L\cos S - \cos L\sin S\cos Z, \qquad (56)$$

and the latitude L' is hence expressed in terms of known quantities.

In the same triangle, using the theorem that the sines of the sides PB and AB are proportional to the sines of the opposite angles,

$$\sin(90° - L')/\sin S = \sin(180° - Z)/\sin(M' - M),$$

or

$$\sin(M' - M) = \sin S\sin Z/\cos L', \qquad (56)'$$

from which the longitude M' can be computed after the latitude L' has been found.

To deduce Z' one of the formulas known as Napier's analogies will be most convenient in numerical work, namely,

$$\tan\tfrac{1}{2}(A+B)/\cot\tfrac{1}{2}C = \cos\tfrac{1}{2}(PB - PA)\cos(PB + PA),$$

and, reducing this to the notation in hand, it becomes

$$\cot\tfrac{1}{2}(Z' - Z) = \tan\tfrac{1}{2}(M' - M)\sin\tfrac{1}{2}(L + L')/\cos\tfrac{1}{2}(L' - L), \qquad (56)''$$

from which the back azimuth Z' can be computed.

These formulas apply to a spherical arc of any size on any sphere. For example, let $L = 40°\ 45'$ and $M = 73°\ 58'$, these being values for New York City, and let it be required to find L', M', and Z', for a point whose angular distance is $S = 35°$ and whose azimuth at New York is $Z = 90°$. From formulas (56) and (56)′

$$\log \sin L' = \bar{1}.72812, \qquad L' = 32° \; 19',$$
$$\log \sin (M' - M) = \bar{1}.83172, \qquad M' = 116° \; 43',$$

which indicates that the point is located in the vicinity of San Diego, California. To find Z' the formula (56)″ gives

$$\log \cot\tfrac{1}{2}(Z - Z') = \bar{1}.36852, \qquad Z' = 243° \; 42',$$

which shows that if a great circle be drawn between the two places the direction of this is due west at New York, but at San Diego its direction is N. 64° E. As the earth is not a sphere, these results may be a degree or more in error.

When a line AB runs due north its azimuth Z is 180°; then (57) reduces to $L' = L + S$ and (57)′ gives $M' = M$, while (57)″ shows that $Z' = Z + 180° = 0°$. If it runs due south Z is 0°; then (57) gives $L' = L - S$ and (57)′ gives $M' = M$, while (57)″ shows that $Z' = Z + 180° = 180°$. If the two points are on the same parallel of latitude, then $L' = L$ and S is their angular distance on a great circle.

For most geodetic triangles the lengths of the sides are so small compared with the radius of curvature R that it is sufficient to take $\cos S = 1$ and $\sin S = l/R$, where l is the length of the arc or chord joining A and B. Then the above formulas may be directly applied to such triangles, and in Art. 64 it will be shown how they are further simplified.

Prob. 56. Given $L = 40° \; 36' \; 22''.452$, $M = 75° \; 22' \; 51''.150$, $Z = 193° \; 56' \; 28''.1$, and $l = 1726.60$ meters. Taking $R = 6371$ kilometers, compute latitude L', longitude M', and azimuth Z'.

CHAPTER VII.

SPHEROIDAL GEODESY.

57. PROPERTIES OF THE ELLIPSE.

Since an oblate spheroid is generated by the revolution of an ellipse about its minor axis, the equator and all the sections of the spheroid parallel to the equator are circles, and all sections made by planes passing through the axis of revolution are equal ellipses. Let a and b represent the lengths of the semi-major and semi-minor axes of this meridian ellipse, which are the same as the semi-equatorial and semi-polar diameters of the spheroid; when the values of a and b have been found all the other dimensions of the ellipse and the spheroid become known. It is necessary first to deduce several equations expressing the properties of the ellipse, and then by discussing them in connection with the results of measurements of meridian arcs the form and size of the spheroid is to be found.

The eccentricity of an ellipse is the ratio of the distance between focus and center to the semi-major axis, and the ellipticity is the ratio of the flattening of one pole to the semi-major axis. Let e be the eccentricity and f be the ellipticity, then

$$e = \frac{\sqrt{a^2 - b^2}}{a}, \qquad f = \frac{a - b}{a}.$$

The relation between these two fractions is

$$e = \sqrt{2f - f^2}, \qquad f = 1 - \sqrt{1 - e^2},$$

and b may be expressed in terms of e and f in two ways,

$$b = a\sqrt{1 - e^2}, \qquad b = a(1 - f).$$

Thus two quantities determine an ellipse; those generally used are a and e, and when these have been found b and f are known.

The equation of the ellipse referred to the rectangular axes QQ and PP is $a^2y^2 + b^2x^2 = a^2b^2$, in which y and x are the ordinate and abscissa of any point A. Let L be the latitude of A, that is, the angle ABQ, and let it be required to find

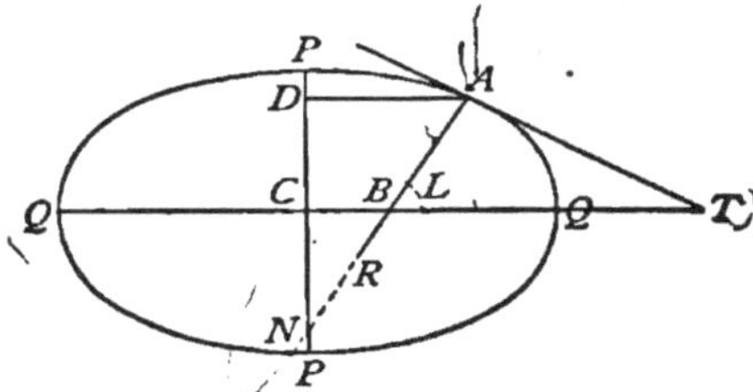

the relation between x and L. At A draw the tangent AT, and, since the angle ATB is $90° - L$,

$$\tan(90° + L) = \frac{dy}{dx} = -\frac{b^2x}{a^2y},$$

whence

$$y = \frac{b^2x}{a^2}\cot L.$$

Inserting this in the rectangular equation of the ellipse, and replacing b^2 by its value $a^2(1 - e^2)$, there is found, after reduction,

$$x = a\cos L/(1 - e^2\sin^2 L)^{\frac{1}{2}}, \qquad (57)$$

which is an equation of the ellipse in terms of the variables x and L. If $L = 0°$ then $x = a$, and if $L = 90°$ then $x = 0$.

To find the radius of curvature of the ellipse at A the rectangular equation of the ellipse may be written in the form

$$y = \frac{b}{a}(a^2 - x^2)^{\frac{1}{2}},$$

and the first and second derivatives of y with respect to x are

$$\frac{dy}{dx} = -\frac{bx}{a(a^2 - x^2)^{\frac{1}{2}}}, \qquad \frac{d^2y}{dx^2} = -\frac{ab}{(a^2 - x^2)^{\frac{3}{2}}},$$

and then, replacing b^2 by $a^2(1 - e^2)$ and x by its value from (57),

$$R_1 = \left(1 + \frac{dy^2}{dx^2}\right)^{\frac{3}{2}} \Big/ \frac{d^2y}{dx^2} = a(1 - e^2)/(1 - e^2 \sin^2 L)^{\frac{3}{2}}, \quad (57)'$$

which is the required radius of curvature. If $L = 0°$ then $R_1 = b^2/a$; if $L = 90°$ then $R_1 = a^2/b$.

To find the length of an arc of the ellipse the differential element $R_1 dL$ is to be integrated between the limits 0 and L. This furnishes an elliptic integral which cannot be evaluated except by a series; thus

$$l = a(1 - e^2)\int_0^L (1 + \tfrac{3}{2}e^2 \sin^2 L + \tfrac{15}{8}e^4 \sin^4 L + \ldots)dL,$$

the integration of which gives

$$l = a(1 - e^2)[(1 + \tfrac{3}{4}e^2 + \tfrac{45}{64}e^4 + \ldots)L - (\tfrac{3}{8}e^2 + \tfrac{15}{32}e^4 + \ldots)\sin 2L + (\tfrac{15}{256}e^4 + \ldots)\sin 4L - \ldots] \quad (57)''$$

for the length of a meridian arc from 0° to $L°$. If $L = 90° = \frac{1}{2}\pi$, then

$$l_q = \tfrac{1}{2}\pi a(1 - \tfrac{1}{4}e^2 - \tfrac{3}{64}e^4 - \tfrac{5}{256}e^6 - \ldots),$$

which is the length of a quadrant of the ellipse.

Prob. 57. Given $e = 0.082271$, and $a = 6\,378\,206$ meters, to compute the length of the quadrant to the nearest meter.

58. Discussion of Meridian Arcs.

Since a spheroid is determined by the two elements a and e of the generating ellipse two equations are required to find their values. These may be established by the discussion of two meridian arcs in different latitudes. Let l_1 and l_2 be their lengths, θ_1 and θ_2 their amplitudes or the number of degrees of latitude between their northern and southern ends, L_1 and L_2 the latitudes of their middle points, and R_1 and R_2

the radii of curvature at these points. Regarding these arcs as arcs of circles, their radii of curvature are

$$R_1 = \frac{180}{\pi}\frac{l_1}{\theta_1}, \qquad R_2 = \frac{180}{\pi}\frac{l_2}{\theta_2},$$

but considering the middle points as lying upon the circumference of an ellipse their radii, as given by (57)′, are

$$R_1 = \frac{a(1 - e^2)}{(1 - e^2 \sin^2 L_1)^{\frac{3}{2}}}, \qquad R_2 = \frac{a(1 - e^2)}{(1 - e^2 \sin^2 L_2)^{\frac{3}{2}}}.$$

Equating the values of R_1 and also the values of R_2 there are found two equations whose solution gives

$$e^2 = \frac{1 - u}{\sin^2 L_2 - u \sin^2 L_1}, \quad \text{where} \quad u = \left(\frac{l_1\theta_2}{l_2\theta_1}\right)^{\frac{2}{3}},$$

$$a = \frac{180 l_1(1 - e^2 \sin^2 L_1)^{\frac{3}{2}}}{\pi\theta_1(1 - e^2)} = \frac{180 l_2(1 - e^2 \sin^2 L_2)^{\frac{3}{2}}}{\pi\theta_2(1 - e^2)}, \tag{59}$$

by which the eccentricity of the ellipse and its major axis may be computed from the data of two measured meridian arcs.

It is plain that these elements will be most accurately determined when one arc is as near the pole as possible while the other is at the equator. These conditions exist in the Lapland and Peruvian arcs (Art. 51), the results of which became known about 1745. The data for these arcs are as follows:

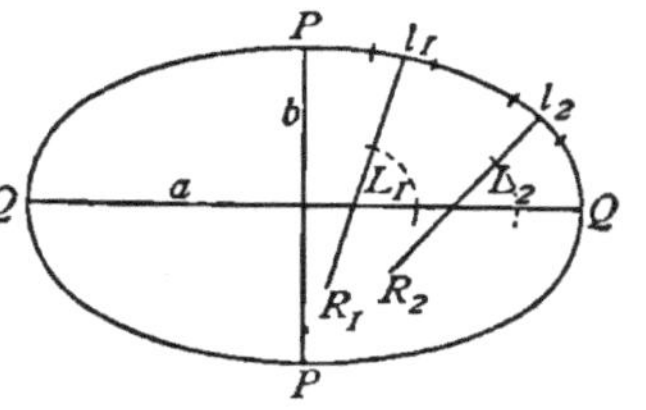

Lapland Arc.	Peruvian Arc.
$l_1 =$ 92 777.98 toises	$l_2 =$ 176 875.5 toises
$\theta_1 =$ 1° 37′ 19″.57	$\theta_2 =$ 3° 07′ 03″.46
$L_1 = +$ 66° 20′ 10″.05	$L_2 = -$ 1° 31′ 00″.34

Reducing the amplitudes to degrees, and substituting in (59), there results

$$e^2 = 0.00643506, \quad \text{whence} \quad e = 0.080219,$$

and then substituting the value of e^2 in either of the values of a there is found $a = 3\,271\,652$ toises. These two values completely determine the ellipse and the oblate spheroid generated by it. Then, from the expressions of the last Article, $f = 0.003223$, and the length of the quadrant is 5 130 817 toises, or 10 000 150 meters.

It is often customary to state the values of e and f as vulgar fractions, since thus a clearer idea of the oblateness of the spheroid is presented. For this case the rough values are

$$e = \frac{1}{12.5}, \qquad f = \frac{1}{310.3},$$

or the distance of the focus of the ellipse from the center is $\frac{1}{12.5}$th and the flattening at one of the poles is $\frac{1}{310.3}$th of the equatorial radius.

About the year 1745 the results of the surveys instituted by the French Academy became known; these have been given in Art. 51 in toises. The length of one degree of latitude is l/θ, if θ be in degrees, and thus these data give everything necessary for computing u, e, and a from the above formulas. From these three arcs three computations were made by the above method, and these gave results about as follows for the ellipticity of the spheroid:

From Lapland and French arcs, $f = \frac{1}{146}$,
From Lapland and Peruvian arcs, $f = \frac{1}{310}$,
From French and Peruvian arcs, $f = \frac{1}{334}$.

Now, if the earth be a spheroid of revolution, and if the measurements be precise, these values of the ellipticity should be the same. Since, however, they disagree the conclusion was easy to make either that the assumption of the spheroid was incorrect or that the surveys were lacking in precision.

After the year 1750, when the results of the Lapland and Peruvian arcs had become known, great interest was manifested in securing additional data by the measurement of other meridian arcs in order to determine whether or not the earth was a true ellipsoid of revolution. The following table

MERIDIAN ARCS.

No.	Locality of Arc.	Middle Latitude.	Length of One Degree.
			Toises.
1	Lapland,	+ 66° 20′	57 405
2	Holland,	+ 52 04	57 145
3	France,	+ 49 23	57 074⅓
4	Austria,	+ 48 43	57 086
5	France,	+ 45 43	57 034
6	Italy,	+ 43 01	56 979
7	Pennsylvania,	+ 39 12	56 888
8	Peru,	− 1 34	56 753
9	Cape of Good Hope,	− 33 18	57 037

gives the results of nine arcs which were measured during the eighteenth century and discussed by Laplace in 1799. For this purpose he took the expression for the radius of curvature given in (57)′, developed it by the binomial formula, and divided it by $180/\pi$, thus obtaining

$$d = \frac{\pi a}{180}(1 - e^2)(1 + \tfrac{3}{2}e^2 \sin^2 L + \tfrac{15}{8} \sin^4 L + \ldots)$$

as an expression for the length of one degree of latitude. It thus appeared that the length of a degree could be expressed by

$$d = M + N \sin^2 L + P \sin^4 L + \ldots,$$

in which $M = \pi a(1 - e^2)/180$, $N = \frac{3}{2}e^2 M$, $P = \frac{15}{8} e^2 M, \ldots$, and Laplace in discussing the above data concluded that it was unnecessary to retain the term containing P since its value is small. Accordingly he wrote

$$d = M + N \sin^2 L,$$

and then proceeded to find probable values of M and N from the nine observations of the above table, and from these to deduce the values of a and e.

At that time the Method of Least Squares was unknown, but Laplace wrote the nine observation equations, and then used the two conditions that the algebraic sum of the errors should be zero and that the sum of the same errors all taken positively should be a minimum. He thus obtained two resulting equations from which he found $M = 56\,753$ toises, $N = 613.1$ toises, and accordingly

$$d = 56\,753 + 613.1 \sin^2 L$$

is an empirical formula for the length of one degree. From these values of M and N he found $e^2 = 2N/3M = 0.007202$, and then $f = \frac{1}{278}$.

The last step in Laplace's investigation is the comparison of the observed values of the lengths of the degrees with those computed from his empirical formula. For the Lapland arc, for instance, observation gives $d = 57\,405$ toises, while the formula gives $d = 57\,267$ toises, the difference, or residual error, being 138 toises, a distance equal to nearly 900 feet, or to nearly 9 seconds of latitude. These errors, says Laplace, are so great that they cannot result from the inaccuracies of surveys, and hence it must be concluded that the earth deviates materially from the elliptical figure.

At the beginning of the nineteenth century it was the prevailing opinion among scientists, founded on investigations similar to that of Laplace, that the contradictions in the data derived from meridian arcs, when combined on the hypothesis of an oblate spheroidal surface, could not be attributed to the inaccuracies of surveys, but must be due in part, at last, to deviations of the earth's figure from the assumed form. This conclusion, although founded on data furnished by surveys that would nowadays be considered rude, has been confirmed by all later investigations, so that it can be laid down as a demonstrated fact that this earth is not an oblate spheroid. Yet it must never be forgotten that the actual deviations from that form are very small when compared with the great size of the globe itself. In some of the practical problems into which the shape of the earth enters it is sufficient to regard it as a sphere, in many others a spheroid must be used, while in only a few cases is it required to regard the deviation from the spheroidal form. Now it was agreed by all in the early part of the nineteenth century, that for the practical purposes of mathematical geography and geodesy it was highly desirable to determine the elements of an ellipse agreeing as closely as possible with the actual meridian section of the earth, or, in other words, that the most probable spheroid should be deduced from the data of observation. This search after the most probable ellipse resulted in the discovery by Legendre, in 1805, of the Method of Least Squares, and the first problem to which this method was applied was a discussion of the elements of the ellipse resulting from five portions of the French meridian arc.

Important geodetic work was carried on in France and Spain by Delambre and Méchain for determining the length of the meter, which, with the accompanying office work, lasted from 1792 to 1807. The meridian arc embraced an amplitude of nearly ten degrees, and the methods for the measurement of bases and angles were greatly improved, in fact approaching for the first time to modern precision. The results were combined with those of the Peruvian arc to find the eccentricity, and this gave for the ellipticity $\frac{1}{334}$ and for the quadrant 5 130 740 toises. This was equivalent to 10 000 000 meters, since by the French law the meter had been defined to be one ten-millionth part of the quadrant. It is now known that this length of the quadrant is too small by nearly 2 000 meters (Art. 60).

Prob. 58. Explain how the method of least squares is to be applied to the deduction of an empirical formula for the length of one degree from the data in the above table.

59. Plumb-line Deflections.

During the nineteenth century many investigations of the size and shape of the elliptical meridian of the oblate spheroid were made. The most important of these gave the results for the ellipticity and for the length of the quadrant which are stated in the following table:

Dimensions of Spheroids.

Year.	By whom.	Ellipticity.	Quadrant in Meters.
1810	Delambre	1/334	10 000 000
1819	Walbeck	1/302.8	10 000 268
1830	Schmidt	1/297.5	10 000 075
1830	Airy	1/299.3	10 000 976
1841	Bessel	1/299.2	10 000 856
1856	Clarke	1/298.1	10 001 515
1863	Pratt	1/295.3	10 001 924
1866	Clarke	1/295	10 001 887
1868	Fischer	1/288.5	10 001 714
1878	Jordan	1/286.5	10 000 681
1880	Clarke	1/293.5	10 001 869

Most of these results were determined by a discussion of the data of several meridian arcs by the Method of Least Squares, and a brief explanation is now to be given as to how such computations are made.

The principle of the Method of Least Squares (Art. 3) requires that the sum of the squares of the errors of observation shall be rendered a minimum in order to give the most probable values of the observed quantities. The first inquiry then is as to where the errors of observation in a meridian arc lie; are they in the linear distance l or in the angular amplitude θ? The error in a linear distance that is computed from a good triangulation is known to be very small, say less than $\frac{1}{200000}$th part of its length (Art. 26). The error in an observed latitude found by the zenith-telescope method cannot exceed half a second (Art. 49). Neither of these errors can account for even a small part of the discrepancy that is found between the observed and computed length of a degree of latitude (Art. 59).

Early in the nineteenth century it was suspected that the cause of these discrepancies was due to deflections of the plumb lines from the normal to the spheroid. To illustrate let the sketch represent a very small part of a meridian section of the earth. *O* is the ocean, *M* a mountain, and *A* a latitude station between them; *ece* is a part of the meridian ellipse coinciding with the ocean surface; *Ac* represents the normal to the ellipse, and *Ah*, perpendicular to *Ac*, the true level for the station *A*. Now owing to the attraction of the mountain *M*, the plumb line is drawn southward from the normal to the position *AC*, and the apparent level is depressed to *AH*. If *AP* be parallel to the earth's axis, and hence pointing toward the pole, the angle *PAh* is the latitude of *A* for the spheroid *ece*; but as the instrument at *A* can only be set for the level *AH*, the observed latitude is *PAH*, which is greater than the former by the angle *hAH*. These differences or errors are usually not large, rarely exceeding ten seconds, yet since a single second of latitude corresponds to about 31 meters or 101 feet, it is evident that the error due to these plumb-line deflections may be very great compared with any accidental error in the measured length of the meridian arc.

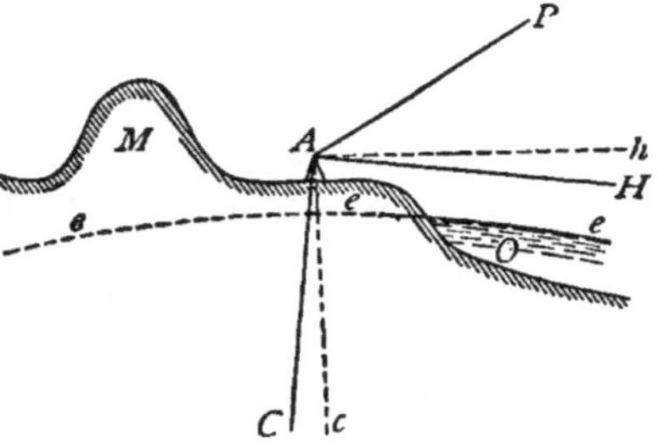

It is not necessary, of course, that there should be any plumb bob on an instrument for determining latitude, but whatever affects a plumb line affects the level bubble by which the graduated limb is made horizontal. Even when a sextant is used the effect of gravity upon the mercury in the artificial horizon may make its surface deviate from parallelism to the tangent plane of the spheroid. Thus, the term plumb-line deflection means really the elevation or depression of the astronomical horizon with reference to the plane tangent to the spheroid. Astronomical latitude is determined with reference to a vertical line at the place of observation (Art. 40), but geodetic latitude is with reference to a normal to a spheroid at that point, and the difference of these is called the plumb-line deflection. A plumb-line direction and an astronomical latitude are real things, but a normal to a spheroid and the corresponding geodetic latitude are artificial things, and hence a plumb-line deviation depends upon the particular spheroid to which it is referred.

In deducing the elements of an ellipse from the data of meridian arcs, the lengths are hence to be taken as without error, and the sum of the squares of the errors in the latitudes is to be made a minimum. For this purpose let l be the length of an arc and θ its amplitude in seconds; the radius of curvature, regarding it as an arc of a circle, is $206\,265l/\theta$, and equating this to the expression for the radius of curvature given by (57)′ there is found

$$\theta = 206\,265l(1 - e^2 \sin^2 L)^{\frac{3}{2}}/a(1 - e^2),$$

in which L is the middle latitude. Now if L' and L'' be the latitudes of the north and south ends of the arc, this expression becomes, after developing the parenthesis and neglecting powers of e higher than the second,

$$L' - L'' = \frac{206\,265l}{a} + 206\,265l(1 - \tfrac{3}{2}\sin^2 L)\frac{e^2}{a}.$$

This equation contains the two elements a and e^2 whose values are to be found, while the other quantities are the data of the meridian arc. Now let v_1 and v_2 be corrections to be applied to the observed latitudes L' and L'', these being the plumb-line deviations from the normals to the spheroid; also let x be a correction to be applied to an assumed value of a, and y a correction to be applied to an assumed value of e^2. Then this equation may be put into the form

$$v_1 - v_2 = mx + ny + p,$$

where m, n, and p are known functions of the observed quantities. Now if there be three meridian arcs, each having two latitude stations, there will be six plumb-line deviations; thus for the first arc the two corrections may be written

$$v_1 = v_2 + mx + ny + p,$$
$$v_2 = v_2,$$

and similarly for the other arcs. If the left members be made zero these are the six observation equations which contain the five unknown quantities v_2, v_4, v_6, x, and y. The normal equations are now formed, and their solution gives the most probable values of x and y, from which those of a and e^2 are found, and also the most probable values of the plumb-line deviations.

Such is a brief outline of the process of determining the size of the earth from several measured meridian arcs. In practice the numerical work is abbreviated by using the method of correlates (Art. 78), but even then is very lengthy, several weeks being required to form and solve the normal equations when many arcs are used. Each arc, moreover, generally has several latitude stations, so that the number of observation equations is more than twice as many as there are meridian arcs. The spheroid thus deduced is the most probable one that can be derived from the given data, for the sum of the squares of the errors in the latitudes has been made a minimum.

Prob. 59. Consult Clarke's investigation of 1866 in Comparison of Standards, published by the British Ordnance Survey, and ascertain the number of meridian arcs used, the number of normal equations, and the greatest values of the plumb-line deflections.

60. DIMENSIONS OF THE SPHEROID.

The most important investigations for the determination of the size of the spheroid are those made by Bessel in 1841 and by Clarke in 1866. The data employed by Bessel included ten meridian arcs, namely, one in each of the countries Lapland, Denmark, England, France, and Peru, two in Germany, and two in India. The sum of the amplitudes of these arcs is about 50.5 degrees, and they include 38 latitude stations. In the manner briefly described above, there were written 38 observation equations, from which 12 normal equations containing 12 unknown quantities were deduced. The solution of these gave the elements of the meridian ellipse, and also the residual errors in the latitudes due to the deflections of the plumb lines. The greatest of these errors was $6''.45$, and the mean value $2''.64$. The spheroid resulting from this investigation is often called the Bessel spheroid, and the elements of the generating ellipse, Bessel's elements; the values of these will be given below.

In 1866 Clarke, of the British Ordnance Survey, published a valuable discussion, which included a minute comparison of all the standards of measure that had been used in the various countries. The data were derived from six arcs, situated in Russia, Great Britain, France, India, Peru, and South Africa, including 40 latitude stations, and in total embracing an amplitude of over 76°. The mean value of the plumb-line deflections, or latitude errors, was found to be $1''.42$. This investigation is generally regarded as the most important one of the last quarter of a century, and the values derived by it as more precise than those of Bessel. The Clarke spheroid, as it is generally called, has been used in most of the geodetic work done in America since 1880, as it is found to represent the earth's true figure in this continent somewhat better than the spheroid of Bessel.

All the results and computations in the following pages of

this volume will be based on the Clarke spheroid of 1866, but it is well for the student to be acquainted with the Bessel spheroid also, since it is extensively used in Europe. The following table gives the complete elements of the two spheroids, and it will be noted that the spheroid of Bessel is smaller than that of Clarke and also less elliptical or oblate. In order to form an idea of the precision of these results it may be noted that the probable error of Bessel's quadrant is 498 meters and that of Clarke's quadrant slightly less.

ELEMENTS OF THE SPHEROID.

		Bessel, 1841.	Clarke, 1866.
Semi-major axis a	meters	6 377 397	6 378 278
	feet	20 923 597	20 926 062
Semi-minor axis b	meters	6 356 079	6 356 654
	feet	20 853 654	20 855 121
Meridian quadrant in meters		10 000 856	10 001 997
Eccentricity e		0.081 697	0.082 271
e^2		0.00667437	0.00676866
Ellipticity f		$\frac{1}{299.15}$	$\frac{1}{294.98}$

The above values of the axis and quadrant of the Clarke spheroid are expressed in legal linear units of the United States, the meter being defined by the statement that it is $\frac{3937}{3600}$ yards, and it will be noticed that they differ slightly from some values used in the preceding pages. Clarke's results were deduced in feet, and then transformed into meters by the relation that a meter contained 3.2808693 feet, as determined by his comparisons of standards. This quantity has since been found to be too great, and according to present knowledge the legal ratio of the United States is very closely the correct and actual one. In the following pages this legal ratio will be used exclusively, namely, one meter = 3.2808333 feet; or the following logarithmic rules may be employed to change meters into feet:

log meters $+ 0.5159841 =$ log feet,
log kilometers $+ \bar{1}.7933502 =$ log miles.

It is well for the student to keep the first of these rules in the memory, but should it be forgotten, it can be found by referring to the last page of the text of this book.

Prob. 60. Compute the lengths of one second of latitude in feet at the pole and at the equator of the Clarke spheroid, using the lengths of one degree as 111 701 meters and 110 568 meters.

61. Lengths of Meridian and Parallel Arcs.

The elements a and e for the Clarke spheroid, substituted in the equations of Art. 57, furnish practical formulas for numerical work. To find an expression for the length of an arc of the meridian between the latitudes L_1 and L_2 the general expression (57)″ is to be integrated between these limits; then representing the mean latitude $\frac{1}{2}(L_1 + L_2)$ by L and the amplitude $L_2 - L_1$ by θ, it can be put into the form

$$l = 111.133.30\theta - 32\,434.25 \sin\theta \cos 2L + 34.41 \sin 2\theta \cos 4L,$$

in which θ is in degrees and l is in meters. For logarithmic work this may be written in the more convenient way,

$$l = [5.0458443]\theta - [4.5110039] \sin\theta \cos 2L + [1.5368] \sin 2\theta \cos 4L,$$

where the numbers in brackets are the logarithms of the constants in the first formula. For instance, to find the length of the meridian from latitude 45° to the pole, put $L_1 = 45°$, $L_2 = 90°$, whence $\theta = 45°$ and $2L = 135°$; then $l = 5\,017\,160.6$ meters, or 16 460 649 feet.

The length of one degree of the meridian for the latitude L is found by making $\theta = 1°$, and then

$$l = 11\,133.30 - [2.752859] \cos 2L + 0.0796] \cos 4L. \qquad (61)$$

For example, let $L = 37°$, then $l = 110\,976.3$ meters, which is the distance on the meridian from latitude $36\frac{1}{2}°$ to latitude

$37\frac{1}{2}°$. By dividing this by 60 the length of one minute results, and a second division by 60 gives the length of one second.

To find an expression for the length of an arc of the parallel, or an arc of longitude, at the latitude L, let the radius of this circle be called r; then the circumference is $2\pi r$ and the length of one degree is $\pi r/180$. The value of r is given by (57), and by expanding the radical, and substituting the values of a and e for the Clarke spheroid, the length of one degree in meters is

$$l=[5.0469490]\cos L-[1.97562]\cos 3L+[\bar{1}.0751]\cos 5L, \qquad (61)'$$

and the length of θ degrees is then θl. For example, if $L = 89°$ then $l =$ 1 949.35 meters is the length of one degree of longitude, and accordingly the distance around the earth on this parallel is about 693 kilometers, or 430 miles.

A more expeditious method of computation, which is sufficiently accurate for arcs of the meridian less than 3 or 4 degrees, and strictly correct for all arcs of parallels, is by the use of the radius of curvature for the given latitude. Thus if R_1 be the radius of curvature of the meridian at latitude L, then l/R_1 is the angle in radians, or $l = R\theta$ is the length if θ be in radians. Accordingly

$$\begin{aligned} l &= R_1\theta \text{ arc } 1°, && \text{when } \theta \text{ is in degrees,} \\ l &= R_1\theta \text{ arc } 1', && \text{when } \theta \text{ is in minutes,} \qquad (61)'' \\ l &= R_1\theta \text{ arc } 1'', && \text{when } \theta \text{ is in seconds.} \end{aligned}$$

The logarithms of R_1 are found in Table IV and those of arc 1°, arc 1′, and arc 1″ in Table VI. For instance, let it be required to find the length in meters between two points on the same meridian whose latitudes are 39° 18′ 12″.8 and 38° 04′ 15″.2. The mean latitude L is 38° 41′ 13″.8, and for this, by interpolation in the table, $\log R_1$ is found to be 6.8034789. The amplitude θ is 4 437″.6 and $\log\theta$ is 3.6461482, while log arc 1″ is $\bar{6}.6855749$. The sum of these gives 5.1352020, whose corresponding number is 136 522.6, which is the distance in meters on the meridian between the two given latitudes; if the result is desired in feet the addition of the constant (Art. 60) gives 5.6511861, which is the logarithm of 447 905 feet.

The same process applies to obtaining the length of an arc of longitude, the radius r being used instead of R_1. For many rough computations the values of the lengths of arcs given in Tables II and III will enable numerical work to be done without using the above formulas.

Prob. 61. Compute the length of the meridian arc from latitude 45° to the equator, and also the length from the pole to the equator.

62. Normal Sections and Geodesic Lines.

At any point on the spheroid let a tangent plane be drawn and perpendicular to this plane let a line be drawn through the point; this line is called the normal and any plane passing through it cuts from the spheroid a normal section. Of these the meridian section is the most important, and next is the prime vertical section, which is the normal section cut by a plane perpendicular to the meridian. These two sections are called principal normal sections because the properties of all other normal sections can be derived from them. The figure shows a point A on the spheroid, NS being the meridian and WE the prime vertical section, while LL is the parallel of latitude and BB an oblique normal section through the point A.

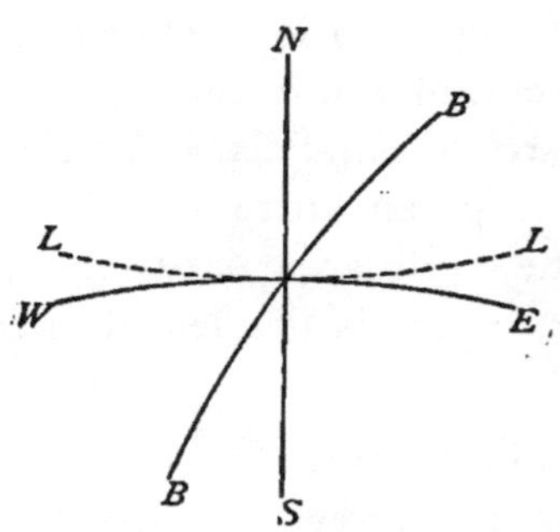

An expression for the radius of curvature of the meridian is given by (57)'. Taking the logarithm of both members and inserting the values of a and e^2 for the Clarke spheroid, it can be put into the form

$$\log R_1 = 6.8039641 - [3.82884] \cos 2L + [0.758] \cos 4L, \quad (62)$$

where the numbers in brackets are logarithms to be added to the logarithms of $\cos 2L$ and $\cos 4L$. A few of the values of $\log R_1$ are given in Table IV at the end of this volume.

The radius of curvature of the prime vertical normal section at its intersection with the meridian is the length of the normal AN from the point A to the intersection with the

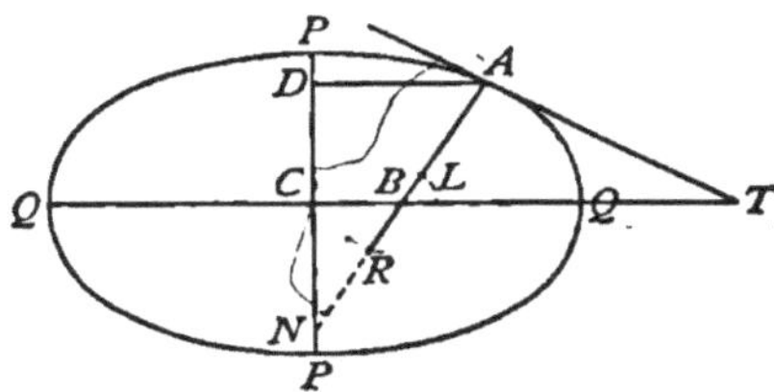

minor axis. It is hence equal to $AD/\cos L$, where AD is the radius of the parallel given by (57). Accordingly

$$R_2 = a(1 - e^2 \sin^2 L)^{-\frac{1}{2}}$$

is the radius of curvature of the prime-vertical normal section. Developing this, there may be deduced

$$\log R_2 = 6.8054402 - [3.35172] \cos 2L + [0.281] \cos 4L, \quad (62)'$$

from which $\log R_2$ may be computed for any given value of L. A few of these values will be found in Table IV.

The radius of curvature of any other normal section at the point A is intermediate in value between R_1 and R_2. If Z be the azimuth of any normal section at the point A, its radius of curvature at that point, as shown in works on the differential calculus, is given by

$$\frac{1}{R} = \frac{\cos^2 Z}{R_1} + \frac{\sin^2 Z}{R_2},$$

which, for numerical work, is better written

$$R = \frac{R_1}{\cos^2 Z} \bigg/ \left(1 + \frac{R_1}{R_2} \tan^2 Z\right).$$

For example, let it be required to find the radius of curvature of a normal section at a point A in latitude 39°, its azimuth being 45°. Taking the logarithms of R_1 and R_2 from Table IV and performing the operations there is found $\log R = 6.8051230$, whence $R = 6.384443$ meters. This formula is useful in reducing base measurements to ocean level (Art. 31).

When an instrument is leveled at a station A and pointed to a second station B, points set out in the line of sight fall in the curve AaB which is cut from the spheroid by the normal section at A. When the instrument is leveled at B and pointed at A, the curve BbA will result, this being cut from the spheroid by the normal section at B. These two curves differ very slightly in azimuth; for a line 100 miles long the difference cannot exceed 0″.1, so that it is of slight importance in common geodetic triangulation. These normal sections are plane curves.

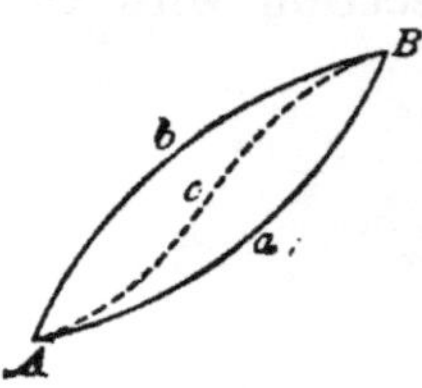

The alignment curve between two stations A and B is a curve traced by starting at A, setting out a point in the direction of B, then moving the instrument to that point, backsighting on A, setting a second point in the direction of B, and so on. The broken line AcB in the figure represents this curve, which is one of double curvature; at any point c the vertical tangent plane to the curve passes through both A and B. The alignment curve is, of course, a shorter path between A and B than that on either of the normal sections.

The shortest line between two stations on the spheroid is called a geodesic line, or simply a geodesic. It is not shown in the above figure, but may be closely represented by a curve of less curvature than AcB and crossing it near c; it is, like the alignment curve, a line of double curvature. The geodesic has the property that the plane containing any element of the curve is normal to the spheroid at that element. The differential equation of the geodesic can be deduced and its properties be studied, but this is not expedient or necessary in an elementary book of this kind, particularly as the line is of no importance in the practical operations of geodesy.

On a sphere the two normal sections and the alignment and geodesic curves between A and B coincide in an arc of a great circle. On a spheroid they also coincide when the two stations are on the same meridian; but in other cases they are separate and distinct. For any two intervisible points on the earth's surface, however, they do not appreciably differ in length, and it is only in the case of the longest lines that a difference in their azimuths can be detected.

Prob. 62. A base line 8046.74 meters long has an azimuth of about 60° and its elevation above ocean level is 1609.35 meters. What is the length of the base reduced to ocean level?

63. Triangles and Areas.

A triangle on the surface of a spheroid has the sum of its three angles greater than two right angles. An exact expression for this spheroidal excess might be established, but, since only triangle sides between intervisible points can be used in geodesy, it is always sufficiently accurate to regard these points as lying on the surface of a sphere osculatory to the spheroid at the middle point of the triangle. The radius of this osculatory sphere is $\sqrt{R_1R_2}$, where R_1 and R_2 are the radii of curvature of the two principal normal sections through the point. Accordingly the formula (55) becomes

$$\text{Excess in seconds} = 206\,265 \text{ area}/R_1R_2, \qquad (63)$$

in which $\log R_1$ and $\log R_2$ may be computed from (62) and (62)′ or be taken directly from Table IV, while the logarithm of 206 265 is found in Table VI.

As this computation is one that is frequently required the

quantity $206\,265/R_1R_2$ may be called $2m$ and the logarithms of values of m for different latitudes be tabulated as is done in the last column of Table IV. Thus the practical formula for the computation of the excess in seconds is

$$\text{Excess} = m \,.\, 2 \text{ Area} = m \,.\, ab \sin C, \qquad (63)'$$

in which $ab \sin C$ is double the area of the triangle, a and b being two sides and C their included angle; here the area must be in square meters, or a and b must be in meters. For instance, let it be required to compute the spherical excess for a triangle whose area is 197 square kilometers, the latitude of its middle point being $37\frac{1}{2}°$. From Table IV the logarithm of m for the given latitude is taken and this added to the logarithm of double the area in square meters gives 0.00026 as the logarithm of the excess in seconds, whence excess $= 1''.001$, which differs by only $0''.001$ from the value given by the rough rule of Art. 55.

Among the many interesting questions relating to the spheroid is that of the areas of zones and the areas of trapezoids bounded by meridians and parallels. The differential expression for the area of a zone is $2\pi r R_1 dL$, where r is the radius of the parallel and R_1 the radius of curvature of the meridian at the latitude L. The values of r and R_1 are given by (57) and (57)', and thus is found

$$A = 2\pi a^2(1 - e^2)\int(1 - e^2 \sin^2 L)^{-2} \cos L \, dL,$$

which, when integrated between the limits L_1 and L_2, gives the area of the zone between those latitudes. The integral contains a hyperbolic or logarithmic function and hence is rather tedious in computation, but tables have been made giving its values. Among the best of these is Woodward's Geographical Tables, published by the Smithsonian Institution, where the areas of trapezoids bounded by meridians and parallels are given in square miles.

Prob. 63. Prove that the entire surface of the spheroid is expressed by $2\pi a\left(1+\frac{1-e^2}{2e}\log_e\frac{1+e}{1-e}\right)$, and show that this reduces to $4\pi a^2$ for the sphere whose radius is a.

64. LATITUDES, LONGITUDES, AND AZIMUTHS.

The formulas established in Art. 56 for the spherical triangle may be adapted to any practical case arising on the spheroid with any required degree of precision. The problem to be solved is as follows: given the latitude and longitude of a point A, the azimuth of AB and its length, to find the latitude and longitude of B and the azimuth of BA. The notation will be the same as that in Art. 56, the given latitude and longitude being designated by L and M, the given distance and azimuth by l and Z, while the required quantities are L', M', and Z'. Let δL be the difference in latitude $L'-L$, and δM be the difference in longitude $M'-M$; also let δZ be the angle by which the meridian at B deviates from parallelism to that at A, so that $\delta Z = Z' - Z - 180°$. Then when δL, δM, and δZ are known, the required quantities will be given by

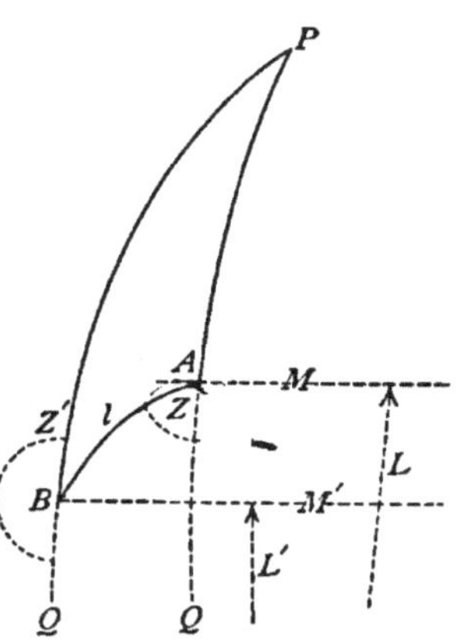

$$L' = L + \delta L,\quad M' = M + \delta M,\quad Z' = Z + 180° + \delta Z. \qquad (64)$$

The problem now is to find formulas for computing δL, δM, and δZ. The solution here given will be sufficient to furnish the results correctly to thousandths of seconds for all cases when the length l does not exceed about 20 kilometers or 12 miles.

Resuming formula (56), and writing $L+\delta L$ in place of L, it becomes, after developing the first member and dividing by $\cos L$,

$$-\sin\delta L = \sin S \cos Z + (\cos\delta L - \cos S)\tan L.$$

Now as both δL and S are small arcs their sines may be taken as equal to the arcs themselves; and also $\cos S = 1 - \frac{1}{2}S^2$ and similarly for $\cos\delta L$. Accordingly the equation reduces to

$$-\delta L = S\cos Z + \tfrac{1}{2}S^2\tan L - \tfrac{1}{2}(\delta L)^2\tan L.$$

Here the first term of the second member is an approximate value of δL, the second term being small since it contains S^2. Accordingly, δL in the third term may be replaced by $S\cos Z$. Further, since δL and S are in radians the value of S is l/R, if R be the radius of curvature at the locality. Accordingly the equation becomes

$$-\delta L = \frac{l\cos Z}{R} + \frac{l^2\sin^2 Z\tan L}{2R^2}.$$

The next question to be considered is regarding the value of R. With regard to the first term, which is the controlling one, it plainly should be the radius of curvature of the meridian passing through the middle of the arc, but as the latitude of that point is not known it is to be taken as that of the meridian at A. With regard to the second term it is not important, since its value is small, what radius should be taken, and it is customary to take that of the osculatory sphere at A. Now let R_1 and R_2 be the radii of curvature of the principal normal sections at A, the values of these being as given in Art. 62; then the equation becomes

$$-\delta L = \frac{l\cos Z}{R_1} + \frac{l^2\sin^2 Z\tan L}{2R_1R_2}.$$

This value is a close approximation, but it can be rendered closer by adding a term to reduce it to the radius of curvature of the meridian at the middle point of the line l. This term will be $\delta L\dfrac{R_1 - R_m}{R_m}$, where R_m denotes that radius. Introducing the general values of the radius from (57)′ for the latitudes L and L_m, replacing L_m by $L - \frac{1}{2}\delta L$, developing, and neglecting terms containing powers of e higher than the square, the additional term is found to be

$$\delta L \frac{R_1 - R_m}{R_m} = (\delta L)^2 \frac{\frac{3}{2}e^2 \sin L \cos L}{(1 - e^2 \sin^2 L)^{\frac{3}{2}}},$$

and accordingly the final formula for the difference of latitude is

$$- \delta L = l \cos Z \,.\, B + l^2 \sin^2 Z \,.\, C + h^2 \,.\, D, \qquad (64)'$$

in which h denotes the value of δL as found from the first and second terms, and in which the letters B, C, and D are factors depending only on the dimensions of the spheroid and on the given latitude L. In order that δL may be found in seconds the above expressions for the constants are to be multiplied by the number of seconds in a radian, and thus

$$B = \frac{206\,265}{R_1}, \quad C = \frac{206\,265 \tan L}{2R_1R_1}, \quad D = \frac{\frac{3}{2}e^2 \sin L \cos L}{206\,265(1 - e^2 \sin^2 L)^{\frac{3}{2}}}$$

are the final factors which can be computed and tabulated for different values of the latitude L.

In order to find the difference of longitude between A and B, formula (56)′ may be used, $M' - M$ being replaced by δM. Since this is small the sine may be replaced by the arc, giving

$$\delta M = \cos S \sin Z / \cos L'.$$

Here, as before, S may be replaced by l/R and the value of the radius should be that of the prime-vertical normal section through B. Introducing this, and reducing from radians to seconds, it becomes

$$\delta M = \frac{k \sin Z \,.\, A'}{\cos L'}, \quad \text{where} \quad A' = \frac{206\,265}{R_1}, \qquad (64)''$$

in which A' is to be used for the latitude L'.

To find the difference in azimuth δZ, formula (56)″ is resumed, and replacing $Z' - Z$ by $180° + \delta Z$, and $M' - M$ by δM, it becomes

$$-\tan\tfrac{1}{2}\delta Z = \tan\tfrac{1}{2}\delta M \sin\tfrac{1}{2}(L + L')/\cos\tfrac{1}{2}(L' - L).$$

Also, since the differences of azimuth and longitude are very small, their tangents are proportional to the number of seconds in their arcs, and

$$-\delta Z = \delta M \sin\tfrac{1}{2}(L + L')/\cos\tfrac{1}{2}\delta L \qquad (64)'''$$

is final formula for computing the difference in azimuth.

The above formulas were derived by Hilgard in 1846, and together with values of the logarithms of the factors A', B, C, and D, will be found in Appendix No. 7 of the Coast and Geodetic Survey Report for 1884; an abridgment of those tables is given in Table V at the end of this book, the proper change being made for the fact that the ratio of meter of that Appendix to the meter of this book is 1.000011. By the help of these the computations may be expeditiously made, as will be illustrated in Art. 66. When using the formulas in connection with these tables it should be remembered that the distance l must be taken in meters.

As a simple example of one application of the formula for δL let it be required to find the number of seconds in a meridian arc whose length is 11 076.4 meters and whose southern end has the latitude 26°. Here $Z = 180°$, $\cos Z = -1$, whence $-\delta L = -l \cdot B + (lB)^2 D$, and by the use of Table V there is found $+\delta L = 360''.953 - 0''.002$, so that the latitude of the north end of the arc is 26° 10′ 00″.951.

Prob. 64. Given $L = 42°$, $M = 80°$, $l = 1000$ meters, and $Z = 90°$ for the point A. Compute L', M', and Z' for the point B.

CHAPTER VIII.

GEODETIC COORDINATES AND PROJECTIONS.

65. THE COORDINATE SYSTEM.

The system of coordinates used in geodesy is generally the angular one employed in geography, latitudes being estimated north and south from the equator and longitudes east and west from the meridian of Greenwich. In North America both latitudes and longitudes are taken as positive and the signs of the coordinates of a point are hence the same as in the linear system of Art. 1. Thus, if a point is determined to have the latitude 40° 19′ 04″.237 and the longitude 85° 07′ 35″.026, it can be at once located roughly on a small-scale map or be precisely located on a large-scale map upon which the meridians and parallels are accurately drawn in a certain system of map projection.

It is well to keep in mind the approximate rules of Art. 53 regarding the lengths of one degree, one minute, and one second of latitude. One second of latitude being about 31 meters or 101 feet, one-tenth of a second is about 3 meters or 10 feet, one-hundredth of a second is 0.3 meters or 1 foot, and one-thousandth of a second is 0.03 meters or 0.1 feet. Precise geodetic work should hence carry the latitudes to thousandths of a second of angle in order to secure a precision comparable with that of precise plane triangulation.

A second of longitude is nearly the same as that of latitude on the equator, but at any other place it is smaller, a rough rule being that it is equal to a second of latitude multiplied by the cosine of the latitude. Thus at latitude 40½°, since cos40½° is 0.774, the length of one second on the parallel is about 78 feet.

The formulas of Art. 61 furnish expressions by which the lengths of one degree, one minute, and one second of both latitude and longitude can be computed for any given latitude L, and values of some of these will be found in Tables II and III. These are sometimes of service in changing angular differences of latitude and longitude between the stations of a secondary triangulation into linear differences, but a more extended table is necessary in order to make such computations with rapidity.

The sketch below gives a representation of the coordinate system of geodesy, the meridians and parallels being roughly drawn on the polyconic projection method which is explained in Art. 69. Station P is located at latitude 40° 45′ and at longitude 86° 43′, while station P' is located at latitude 40° 36′ and longitude 86° 58′. The straight line, or geodesic, joining the points is very slightly curved in the projection. The azimuth of PP' is about 53° 43′, this being measured from the south around toward the west. The azimuth of $P'P$ is about 233° 33′, this being also measured from the south around through the west and north. Owing to the convergence of the meridians that pass through P and P', the back azimuth of $P'P$ differs by 10′ from the azimuth of PP' plus 180 degrees.

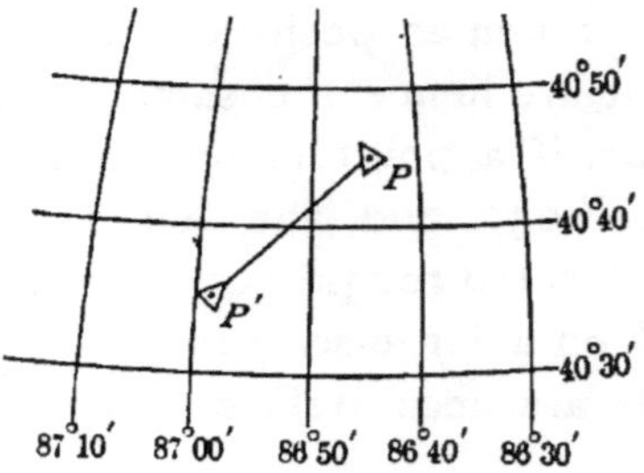

This coordinate system is not a convenient one for the use of local surveyors, but in an area of considerable extent it is a necessary one for the location of points in their relative positions on the spheroid. For an area of moderate size it may be modified in many ways, one of these being the well-known system of the public land surveys of the United

States, while another and more satisfactory system is that of linear rectangular spherical coordinates which is described in Art. 70.

It is well to note again that the latitudes and longitudes used in geodetic work do not generally agree with the latitudes and longitudes obtained by astronomical observations. Thus, if these coordinates be found astronomically for an initial station P together with the azimuth of PP', and if the distance PP' be directly measured or be found by computation from a measured base, then the latitude and longitude of P' and the azimuth of $P'P$ can be computed by the methods of the last Article, and these computed values are called geodetic ones. If further astronomical observations be made at P', the results will generally differ from the geodetic ones owing to the plumb-line deflection at P'. That is to say, the Clarke spheroid passed through P and oriented by the astronomical work done there, has a tangent plane at P' which is not parallel to the astronomical horizon at that point (Art. 59). The differences of latitude and longitude as found by geodetic triangulation are, in fact, always far more precise than those derived from astronomical observations, and it is only by the field operations of geodesy that coordinates of stations can be found so as to form a reliable basis for accurate surveys.

Prob. 65. From the above data for stations P and P' determine roughly, with the help of Tables II and III, the length of the line PP' in miles.

66. *LMZ* Computations.

When the latitude L and longitude M of a station P are given, together with the distance l and azimuth Z from it to a second station P', the latitude L' and longitude M', together with the back azimuth Z', can be computed. The formulas of Art. 64 will determine the latitudes and longitudes correctly to thousandths of a second when the length of the line does not exceed about 20 kilometers or 12 miles, and correctly to hundredths of a second for much longer distances. These formulas will now be exemplified.

Let the given station be one called Bake Oven, whose known latitude is 40° 44′ 54″.109 and whose known longitude is 75° 44′ 02″.222. Let the distance and azimuth to a second station called Packer Spire be 33 932.55 meters and 297° 36′ 49″.42. In the form below these data are seen in italic type, together with the logarithm of l in two places; then log sin$Z$ and log cos$Z$ are found and put in their places, while the logarithms of l^2 and sin^2Z are found by doubling those of l and sin$Z$, and the logarithms of the factors B, C, and D are taken from Table V. By addition the logarithm of l cos$Z \cdot B$, or h, is found, and its double is the logarithm of h^2. Then the logarithms of the second and third terms are found, and thus the final value of $-\ \delta L$ is 511″.8, from which the latitude of Packer Spire at once results, as also the mean latitude $\frac{1}{2}(L + L')$. The longitude computation is now made as indicated by the formula, the factor A' being taken for the latitude L'. Lastly, δZ is computed, and the back azimuth from Packer Spire to Bake Oven is determined by the rule $Z' = Z + 180° + \delta Z$.

A second or check computation should always be made whenever there is another station that furnishes sufficient data. For this case the latitude of a station called Smith's Gap is 40° 49′ 21″.787 and its longitude 75° 25′ 21″.906, while the distance and azimuth from it to Packer Spire are 24 332.28 meters and 351° 11′ 08″.84. Inserting these data in another form, and carrying out the computations in the same manner, the value of L' will be found to agree within 0″.006, or 0.6 feet, with that of the first computation, while the value of M' will be found to agree within 0″.002. These discrepancies are due to the fact that both lines exceed 20 kilometers in length. The back azimuth from Packer Spire to Smith's Gap is found to be 171° 12′ 52″.29, and the

FORM FOR *LMZ* COMPUTATION.

PACKER SPIRE computed from BAKE OVEN.

$- \delta L = l \cos Z . B + l^2 \sin^2 Z . C + h^2 . D$

$+ \delta M = l \sin Z . A'/\cos L$

$- \delta Z = \delta M \sin \frac{1}{2}(L + L')$

BAKE OVEN

PACKER SPIRE

Z	Bake Oven to Packer Spire	297° 36′ 49″.42
δZ		+ 13 53 .59
180°		180
Z'	Packer Spire to Bake Oven	117° 50′ 43″.01

L	40° 44′ 54″.109	Bake Oven	M	75° 44′ 02″.222
δL	− 08 31 .859	l = 33932.55 *meters*	δM	− 21 18 .904
L'	40° 36′ 22″.250	Packer Spire	M'	75° 22′ 43″.318

l	4.5306166	l^2	9.06123		
$\cos Z$	$\bar{1}$.6660516	$\sin^2 Z$	$\bar{1}$.89496	h^2	5.4149
B	$\bar{2}$.5107900	C	$\bar{9}$.33974	D	$\bar{8}$.3882
h	2.7074642		0.29593		$\bar{3}$.8031
1st term	+ 509″.8755	2d term	1″.9767	3d term	0″.0064
2d and 3d terms	1 .9831				
$- \delta L$	+ 511 .859	l	4.5306166		
		$\sin Z$	$\bar{1}$.9474790		
$\frac{1}{2}\delta L$	04′ 15″.93	A'	$\bar{2}$.5090982	δM	3.1068370
$\frac{1}{2}(L + L)$	40° 40′ 38″.18	$\cos L'(a.c)$	0.1196432	$\sin\frac{1}{2}(L+L')$	$\bar{1}$.8141128
			3.1068370		2.9209498
		$+ \delta M$	− 1278″.904	$- \delta Z$	− 833″.59

difference between this and the back azimuth to Bake Oven is 53° 22′ 09″.28, which furnishes a final check on the work, as this is the value of the spherical angle at Packer Spire.

The best way to carry on these two computations is to enter the data in both, find log sin$Z$ and log cos$Z$ for both at the same time, take out the factors B, C, and D for both, and thus finish the computation of δL in both at the same time. If these values agree, as they should unless the lines are too long, the two computations for δM may be made, and lastly the two for δZ. The signs of δL and δM can in all cases be found from the signs of $\cos Z$ and $\sin Z$, but it will be just as well for the student to determine them from the figure that should always be drawn at the top of each computation sheet.

The above formulas may be used in finding coordinates to tenth of seconds or to single seconds for primary lines of almost any length, but when these are required to thousandths of seconds additional terms are needed. These terms and the form for computation will not be presented in this elementary book, but they may be found in the paper of the Coast and Geodetic Survey cited in Art. 64.

Prob. 66. Using the above latitudes and longitudes of the stations Bake Oven and Smith Gap, and also the data that the distance and azimuth from Bake Oven to Topton are 30 433.63 meters and $351° 48' 49''.11$, and from Smith Gap to Topton are 44 239.59 meters and $29° 54' 17''.84$, make the two LMZ computations for Topton, and check the back azimuths by comparison with the spherical angle at Topton, whose value is $37° 53' 19''.01$.

67. The Inverse LMZ Problem.

When the latitudes and longitudes of two stations are given, it is possible, if they are not too far apart, to compute the length of line joining them and the front and back azimuths of that line. This is readily done in a plane system of coordinates, as illustrated in Art. 24, but in a geodetic system it is more difficult. This is called the inverse LMZ problem, and it will now be shown how the formulas of Art. 64 are applied to its solution.

Since the latitudes L and L' are given, as also the longitudes M and M', the values of δL and δM are known. Then formulas $(64)'$ and $(64)''$ may be written in the form

$$-\,\delta L \;= (l\cos Z)B + (l\sin Z)^2C + (\delta L)^2 . D, \quad (67)$$
$$+\,\delta M = (l\sin Z)A'/\cos L', \quad (67)'$$

in which l and Z are two unknown quantities to be found. From (67)′ the value of $(l\sin Z)$ at once results and this inserted in (67) gives the value of $(l\cos Z)$; then, by dividing the former by the latter $\tan Z$ is found and hence Z. Also dividing $(l\sin Z)$ by $\sin Z$ the value of l results. Lastly δZ is computed by (64)‴ and Z' by (64). The form used in Art. 66 may be advantageously employed in making the computations, as will now be exemplified.

Let the latitudes and longitudes of the stations Smith Gap and Bake Oven be given as stated in the last Article, and let it be required to compute l, Z, and Z'. These, with the resulting values of δL and δM, are first inserted in the form as seen in italic type. From Table V the factors A', B, C, and D are taken. The logarithm of δM is found, then those of $\cos L'$ and A', and accordingly the logarithm of $(l\sin Z)$ results. From δL the logarithm of $(\delta L)^2$ is found, and that of $(l\sin Z)^2$ being also known, the second and third terms of the value of δL are determined, and finally the first term whose logarithm is then known and from which the logarithm of $(l\cos Z)$ results. Then $\tan Z$ is obtained as explained above whence Z is found; then the logarithm of l and its value in meters is determined. Lastly, the computation of δZ is made and the back azimuth Z' is obtained.

A check computation for this case should also be made by changing the order of the stations; thus the values of L and M may be taken for the station Bake Oven and L' and M' for the station Smith's Gap. If the lengths of the lines do not exceed 20 kilometers the values of the lengths and azimuths should exactly agree with those of the first computation. This inverse solution is often advantageous in field work in determining the directions between stations which are not connected by a triangle side.

INVERSE *LMZ* COMPUTATION.

$$-\delta L = (l\cos Z)B + (l\sin Z)^2C + (\delta L)^2D$$
$$+\delta M = (l\sin Z)A'/\cos L'$$
$$-\delta Z = \delta M \sin\tfrac{1}{2}(L+L')$$

SMITHS GAP
Z
Z'
BAKE OVEN

Z	Smith's Gap to Bake Oven	72° 39′ 07″.24
δZ		− 12 11 .82
180″		180
Z'	Bake Oven to Smith's Gap	252° 26′ 55″.42

L	40° 49′ 21″.787	Smith's Gap	M	75° 25′ 21″.906
δL	− 04 27 .678	l = 27535.63 meters	δM	− 18 40 .316
L'	40° 44′ 54″.109	Bake Oven	M'	75° 44′ 02″.222

$(l \cos Z)$	3.9143619	$(l^2 \cos^2 Z)$	8.83935	$(\delta L)^2$	4.8552
B	$\bar{2}.5107842$	C	$\bar{9}.34087$	D	$\bar{8}.3884$
h	2.4251461		0.18022		$\bar{8}.2436$
1st term	− 266″.162	2d term	+ 1.5143	3d term	+ 0″.0018
2d and 3d terms	+ 1 .516				
$-\delta L$	− 267″.678	$(l \sin Z)$	4.4196765		
		A'	$\bar{2}.5090946$	δM	3.049340
$\log \delta L$	2.42762	$\cos L'(a.c)$	0.1205694	$\sin\frac{1}{2}(L+L')$	$\bar{1}.815066$
$\frac{1}{2}(L+L')$	40° 47′ 07″.9				
			3.0493405		2.864406
		$+\delta M$	− 1120″.316	$-\delta Z$	− 731″.82

$(l \sin Z)$	4.4196765	$(l \sin Z)$	4.4196765
$(l \cos Z)$	3.9143619	$\sin Z$	$\bar{1}.9797815$
$\tan Z$	0.5053146	l	4.4398950

Prob. 67.-Make the inverse *LMZ* computation for the above data, taking L and M for the station Bake Oven and L' and M' for the station Smith's Gap.

68. Map Projections.

As a surface of double curvature cannot be developed on a plane it is impossible to devise any method of representing a large area on a map without some distortion. The method of orthographic projection is perfectly satisfactory for a small area, but when applied to the whole earth, or even to a large county, the features near the edges of the map are crowded together so as to appear unnatural. For instance, in the lower diagram of the figure, which shows an orthographic projection of the northern hemisphere on the plane of the equator, it is seen that the distance between parallels of latitude near the outer parts of the map is much less than near the central part; in the upper diagram, which is an orthographic projection on the plane of one of the meridians, a similar distortion is also observed.

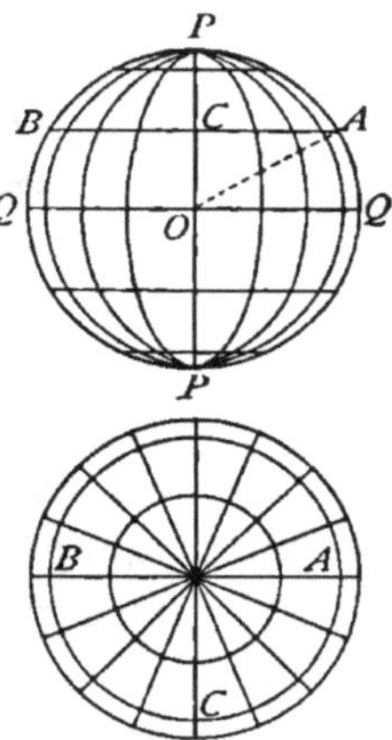

A projection devised by Flamsted to avoid this distortion consists in dividing the central meridian *NS* into parts proportional to the distance between the parallels, and through these points drawing straight lines to represent those parallels. Each parallel is then divided into the same number of equal parts, and the meridians are drawn through these points of division. In this method each trapezoid has the same area and consequently much of the distortion of the orthographic method is avoided.

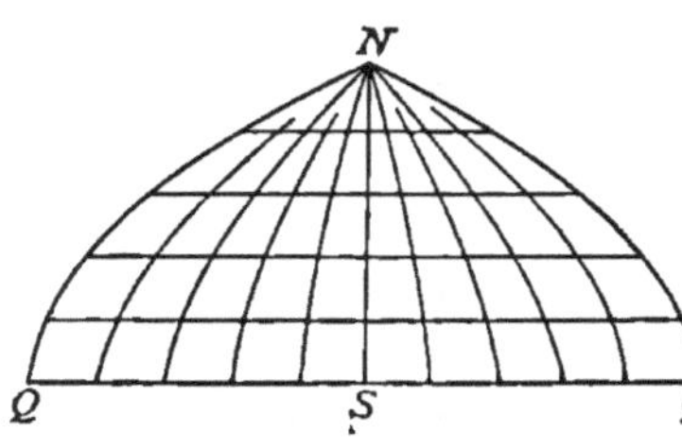

A projection devised by Bonne is constructed in a similar manner to that of Flamsted, except that the parallels are concentric circles. The center of these circles is in the middle meridian and at a distance of a cot$L$ from the middle parallel whose latitude is L; thus in representing half of the hemisphere the radius of the middle parallel is equal to the

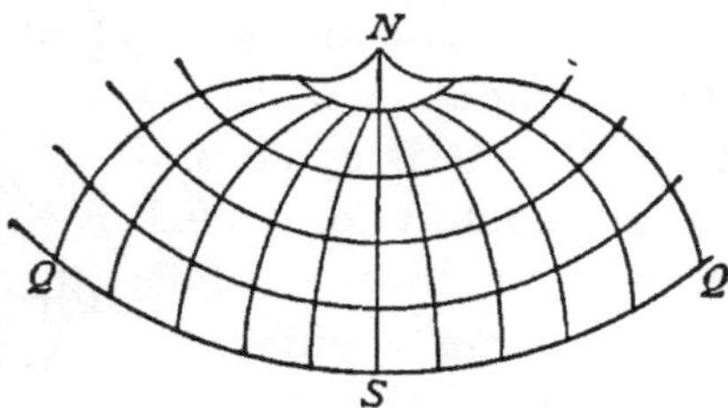

equatorial radius a, and the radius of any other parallel is $a \pm d$, where d is its distance from the middle parallel. This method gives trapezoids of equal area and an orthographic projection along the middle meridian and parallel, but the shape becomes rather awkward where a large area is represented.

Mercator's projection of the surface of a sphere is made by projecting the parallels upon a circumscribing cylinder by lines drawn from the center of the sphere, and then developing the cylindrical surface. If QPQ be a meridian section of the spheroid and d any point upon it, then d is projected at D on the cylinder, and thus the parallel through d is projected upon the cylinder in a circle whose diameter is $D'D$. The cylinder being developed on a plane tangent to the cylinder, the circle $D'D$ rolls out into the straight line D_1D_1, while the equator rolls out in its true length on the line Q_1Q_1. If L be the latitude of any parallel and R the radius of the sphere, the distance of the parallel from the equator on the development is R tan$L$. Thus the distances between parallels increases toward the poles and the poles themselves cannot be shown on this projection. The equator being divided into equal parts, representing degrees of longitude, the meridians are drawn parallel to each other, and accordingly one degree of longitude has the same length on all parts of the map.

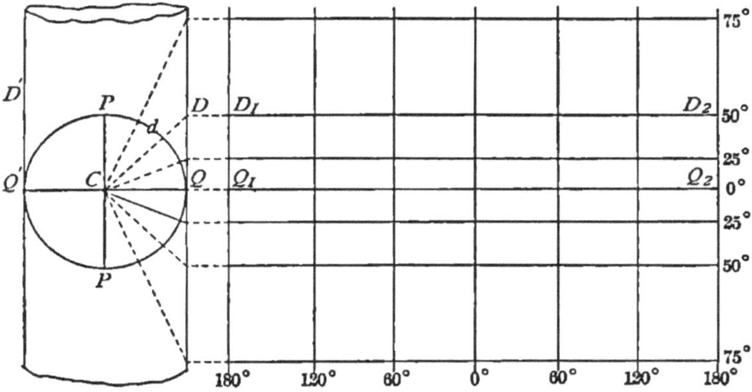

In the case of the spheroid the same principles apply for the construction of a Mercator projection, but the distance of any parallel from the equator will be

$$d = a \tan L - ae^2 \sin L(1 - e^2 \sin^2 L)^{-\frac{1}{2}},$$

as may be proved by finding an expression for BQ in the figure of Art. 57 and multiplying it by $\tan L$, since in that figure the cylinder is to be made tangent to the spheroid at QQ and BA is to be produced to meet it. This projection is a favorite one with navigators, since the course of a ship is plotted on the map in a straight line as long as it runs on the same true bearing. It is, however, an inconvenient projection for plotting distances. Along the equator the projection is orthographic and a distance may be laid off in its true length. At any latitude L a distance l has the length $l/\cos^2 L$ when laid off along a meridian and the length $l/\cos L$ when laid off along a parallel. In the polar regions the distortion is so great that the projection is unsatisfactory north of latitude 60 degrees.

Prob. 68 Deduce the algebraic expressions in the last paragraph, and state something about the life of Mercator.

69. The Polyconic Projection.

The map projection that is used exclusively in geodetic work is one in which each parallel circle of latitude is developed on a conical surface, there being as many cones as there are parallels. In the figure let A be any point on the spheroid

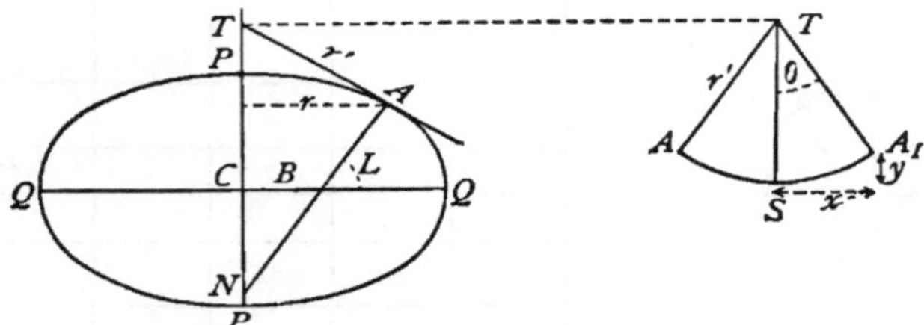

whose latitude is L and let r be the radius of its parallel of latitude whose value is given by (57). Let a tangent cone be drawn touching the spheroid at this circle of latitude; in the figure AT is one element of this cone, its vertex being at T where AT meets the polar axis. If this cone be developed, as in the second diagram, the parallel of latitude rolls out into a circle ASA_1, whose radius is the same as TA. In the same manner a tangent cone may be rolled out for any other circle of latitude, its radius being different from that in the first case.

For large-scale maps the radius TA, or r', is so long that it is impracticable to describe the circle ASA_1 with the compass, and hence it is usually constructed by finding the abscissas x and the ordinates y with respect to a point S on the central meridian TS. Since the angle ATN is the same as L the value of r' is $r/\sin L$. Now if it be desired that SA_1 shall correspond to n degrees of longitude, the length of this parallel arc is rn, but in the projection its length is $r'\theta$, where θ is the angle A_1TS. Therefore, equating these, the value of θ is given by

$$\theta = n \sin L. \qquad (69)$$

After θ is found the abscissa and ordinate result from

$$x = r' \sin\theta, \qquad y = r'(1 - \cos\theta) = 2r' \sin^2\tfrac{1}{2}\theta, \quad (69)'$$

in which r' has the value stated above. In numerical work it is preferable to express r' in terms of R_2, the radius of curvature of the prime-vertical section at A (Art. 62), whose value is the same as that of the line AN. Since $R_2 = r \tan L$, it follows that

$$r' = R_2 \cot L, \qquad (69)''$$

and thus r' can be easily computed by the help of Table IV.

For example, at latitude 40° let it be required to find the values of x and y, for 2°, 4° and 6° of longitude. From Table IV the logarithm of R_2 is 6.8053115, and accordingly

for $n =$	2°	4°	6°
$\theta =$	1° 17′ 08″.07	2° 34′ 16″.14	3° 51′ 24″.21
$x =$	170 780	341 475	511 996 meters
$y =$	1 916	7 663	17 238 meters

and by these three points may be located on each side of the central point S of the developed circle of latitude.

In order to construct a polyconic projection for an area embracing twelve degrees of longitude and eight degrees of latitude, with meridians at intervals of two degrees, five computations like the above are to be made. If the amplitude is from latitude 40° to latitude 48°, a straight line NS is drawn for the central meridian. At S a straight line normal to NS is drawn,

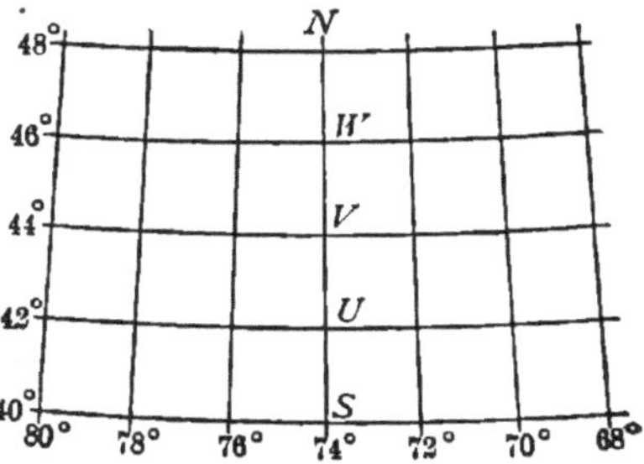

and then six points on the 40° circle are located with the help of the above values of x and y. The distances SU, UV, VW, and WN are next laid off by the help of Table II and through the points thus determined straight lines are drawn normal to NS, and on each of these the values of x and y are laid off as computed for the latitudes 42°, 44°, 46°, and 48°, thus locating six points on each parallel. Through these points curves are drawn and the coordinate system is completed.

Tables of the lengths of arcs of the meridian and of values of the coordinates x and y are indispensable in the construction of polyconic projections. Extended tables in meters may be found in Appendix No. 6 of the Report of the U. S. Coast and Geodetic Survey for 1884, while similar ones for the English system may be found in Woodward's Geographical Tables, published by the U. S. Smithsonian Institution.

The polyconic projection furnishes a system of coordinates that gives an excellent representation of the earth's surface or of any part of it. The parallels and meridians everywhere intersect at right angles. Distances are correctly represented on the central meridian and on all the parallels. The distances on meridians near the borders of the map are a little too long, and it is here that the distortion is greatest. On the whole this method gives the best projection of angles with the least possible distortion of figures.

Prob. 69. Make all the necessary computations and construct a polyconic projection on a scale of 1/2 000 000 for the portion of the spheroid indicated by the above figure.

70. Linear Spherical Coordinates.

For the purposes of local surveys the angular system of coordinates is not a convenient one, as surveyors require all distances to be expressed in linear measures. The system of rectangular linear spherical coordinates, now to be described, is a generalization of the linear method given in Chapter II, and can be applied satisfactorily to a territory of several thousand square miles. This system has been extensively used in Europe, but is little known in America, where precise detailed surveys of large areas have as yet not been made.

Let O be an origin of coordinates at the center of the territory to be covered by the system, NS a central meridian, and R the radius of curvature of the spheroid at O, that is, the radius of a sphere osculatory to the spheroid at the origin. Through any point P_1 let an arc of a great circle of this osculatory sphere be drawn normal to the meridian NS, meeting it in M_1. Then the linear distances OM_1 and M_1P_1 are the linear rectangular spherical coordinates of P_1, and these will be expressed by the letters L_1 and M_1. Similarly, for another point P_2 the latitude L_2 is the distance OM_2, and the longitude M_2 is the distance M_2P_2. Here latitudes are taken as positive when measured northward of O, and longitudes as positive when measured westward from NS.

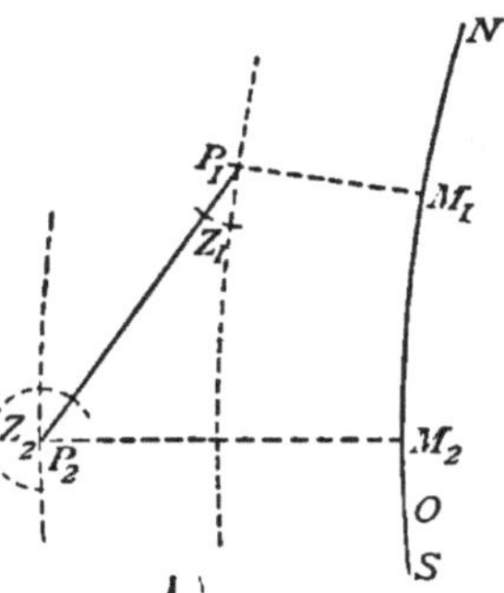

Through P_1 and P_2 let circles be drawn parallel to the central meridian NS, and let the angle which the line P_1P_2 makes with the circle through P_1 be called Z_1. Here Z_1 is not strictly an azimuth, and it is best to call it a direction-angle. Similarly, the direction-angle of P_2P_1 at P_2 is called Z_2. These direction-angles are measured from the south around through the west, north, and east exactly like geodetic azimuths.

Let l be the length of the line P_1P_2, and let also the latitude L_1, the longitude M_1, and the direction-angle Z_1 be given; it is required to find the latitude L_2, the longitude M_2, and the direction-angle Z_2. The solution in the case of a plane is

$$L_2 = L_1 - l \cos Z_1, \quad M_2 = M_1 + l \sin Z_1, \quad Z_2 = Z_1 + 180°,$$

and in the case of the sphere the same expressions will result with the addition of terms of small numerical value which contain the radius of curvature R. The formulas here given are those deduced by Soldern in 1809; the demonstration, though not difficult, will be omitted. First, let m and n be computed from

$$m = l \cos Z_1, \qquad n = l \sin Z_1.$$

Then the required quantities are determined by

$$M_2 = M_1 + n + \frac{m^2 M_1}{2R^2} - \frac{m^2 n}{6R^2},$$

$$L_2 = L_1 - m - \frac{m M_2^2}{2R^2} + \frac{mn^2}{6R^2},$$

$$Z_2 = Z_1 + 180^\circ - \frac{206\,265 mM_1}{R^2} + \frac{206\,265 mn}{2R^2},$$

in which the terms containing R^2 are easily computed by the use of four-place logarithms. Since $R^2 = R_1 R_2$, where R_1 and R_2 are the radii of curvature of the two principal normal sections, the logarithm of R^2 is readily found from Table IV at the end of this book.

For example, suppose such a coordinate system to be used for a region whose middle latitude is 40°. From Table IV the logarithms of the constants are, in meters,

$$\log(1/2R^2) = \overline{13}.0901, \qquad \log(1/6R^2) = \overline{14}.6130,$$
$$\log(206\,265/R^2) = \bar{9}.7055, \quad \log(206\,265/2R^2) = \bar{9}.4045.$$

Now suppose there be given $M_1 = +42\,585.934$ meters, $L_1 = +51\,449.866$ meters, and $Z_1 = 16^\circ\,47'\,09''.38$. Then $m = +26\,343.669$ meters, $n = +7\,946.133$ meters, and

$$M_2 = 42\,585.934 + 7\,946.133 + 0.233 - 0.023 = +50\,532.277 \text{ meters},$$
$$L_2 = 51\,449.866 - 26\,343.669 - 0.123 + 0.007 = +25\,106.081 \text{ meters},$$
$$Z_2 = 16^\circ\,47'\,09''.38 + 180^\circ - 03''.66 + 00''.53 = 196^\circ\,47'\,06''.25,$$

and thus the point P_2 is completely determined.

One of the great advantages of this system is that the difference of the front and back direction-angles of a line differs but little from 180°, and either may be used by a local surveyor to check his topographic work. With geodetic azimuths, on the other hand, the orientation of such local work may be more accurately made at the starting station, but when checking on a second station the large difference in direction is liable to lead to confusion. Undoubtedly the system of linear spherical coordinates must in time come into use in America, and by it or some similar method the results of the geodetic triangulations can be made more generally available for use in precise detailed surveys of large areas. For a fuller account of the system, as also for the method of finding the distance and direction-angle between two stations whose latitudes and longitudes are given, reference is made to Vol. II of Jordan's Handbuch der Vermessungskunde.

Prob. 70. Given $L_1 = + 50\,000$, $M_1 = - 10\,000$, $L_2 = + 60\,000$, and $M_2 = - 20\,000$ meters, to find the distance from P_1 to P_2 and the direction-angles Z_1 and Z_2.

CHAPTER IX.

GEODETIC TRIANGULATION.

71. RECONNAISSANCE.

The first thing to be done in a reconnaissance for selecting the stations of a geodetic triangulation is to make a careful study of all existing maps. Small-scale sketch maps should be prepared, showing the principal watersheds and mountain ranges as far as they are known, and these are to be taken into the field by the reconnaissance party. Such a party consists of two or three men and it is equipped with aneroid barometers, prismatic pocket compasses, binocular field glasses, and photographic cameras, together with apparatus for climbing trees.

Ascending to one of the highest elevations in the region a series of sketches showing the visible horizon and intermediate hill ranges is made. On this are marked the magnetic bearings and the estimated distances to all prominent peaks and gaps. Photographs of the portions of the horizon where it seems probable that stations may be located should also be taken, and the names of all mountain ranges and peaks be ascertained. Then, ascending to another elevation several miles away, a similar series of sketches is made, and after several of these observations the party obtains a fair idea of the topography of the country. The heights of all the positions occupied in this work are to be obtained as closely as can be done with the aneroid.

The results of these operations are to be plotted from day-to day on the sketch maps, the visible horizon as seen at each station being roughly drawn, and the intersection of these horizon lines, together with the observed heights, will give information regarding the approximate positions to be selected for the primary stations. In many cases there are one or two peaks so prominent that no doubt exists as to their availability for stations, while regarding others much additional field work must be done before a final decision can be made. The intervisibility of adjacent stations must of course be insured, and in a prairie country where high towers are to be erected this requires the application of the rules of Art. 37 regarding curvature of the earth and refraction. The primary stations are to be so selected as to secure the best triangle, polygon, or quadrilateral nets to cover the given area under the prescribed conditions of precision and cost, care being taken to avoid angles less than 30 degrees, except in quadrilaterals (Art. 17). As a general rule for primary triangulation the longest possible lines are to be obtained which are consistent with the formation of well-proportioned triangles.

As an example of one of the field computations, suppose that two stations 28 miles apart are 65 and 105 feet, respectively, above the ocean level, and that the highest point between them is on a ridge 10 miles from the first station and 20 feet above ocean level. It is required to find the height of towers at the two stations so that the line of sight may pass 10 feet above the top of the ridge. Assuming that the line of sight is parallel to a tangent to the level surface at the ridge, the combined effect of curvature and refraction is $0.57 \times 10^2 = 57$ feet for the first station and $0.57 \times 18^2 = 185$ feet for the second station. Hence the height of the tower at the first station should be $57 + 30 - 65 = 22$ feet, and at the second station $185 + 30 - 105 = 110$ feet.

Reference may be made to the Reports of the Coast and Geodetic Survey for 1882 and 1885 for a full account of the rules of reconnaissance, and to Final Results of the Triangulation of the New York State Survey (Albany, 1887) for an interesting description of the detailed field work.

After the reconnaissance party has established a few stations a triangulation party may start at work in the measurement of angles. It is the duty of this party to mark the stations, erect the towers and signals, and make the observations of the horizontal and vertical angles. Sometimes the reconnaissance and triangulation work are done by the same party, this method usually saving expense. Base-line measurement and astronomical work are, however, usually done by specially trained parties.

Prob. 71. A station B is 325 feet above A, but between them, at a distance of 15 miles from A and 25 miles from B, is a ridge which is 10 feet above A. If no tower is built at B and one 50 feet high at A, how much above the ridge does the line of sight pass?

72. Stations and Towers.

The marking of a station in a permanent manner is usually done by the first triangulation party which takes the field, the reconnaissance party merely selecting and describing the approximate location. It is believed, however, that if the responsibility of marking the station were assigned to the reconnaissance party, a better location would often be made. The name of the station is usually assigned by the reconnaissance party, and this should be the same as the local name of the peak or ridge on which it is situated.

The stations are marked by bolts set into the rock, or by stone monuments set in the ground. In the latter case it is customary to bury beneath the monument a bottle or crock whose center marks the center of the station. When this is done the knowledge of the bottle or crock should be concealed from the people of the neighborhood, and it should be covered with a large flat stone having a hole drilled in its upper surface. The bottom of this flat stone should be about six inches above the crock, its top about three feet below the surface of the ground, and upon it the foot of the monument may be set. The centers of the underground mark, of the hole in the flat stone, and of the top of the monument should be in the same vertical. Near the top of the monument "U. S." or other appropriate letters should be cut. Detailed instructions regarding the methods of marking stations may be found in the Reports above cited. Reference points should be located on surrounding rocks, or by auxiliary monuments, from which bearings and distances are to be measured to the station. The geodetic surveyor should always make his description of the station clear and full, so that it may be found after the lapse of many years. For this purpose it is well to run a traverse line to the nearest public road, if there is one within a reasonable distance, and erect there a monument which may serve as a starting point for future parties.

A tower is a structure erected over a station for the support of the theodolite and observer. It consists of two parts, an interior tripod to carry the instrument, and an exterior scaffold entirely surrounding the tripod but unconnected with it. The interior tripod is usually made of three posts braced together, while the outside scaffold is a structure like a braced pier having four posts. Rough towers made of timber cut on the spot can be built for about $1.00 per vertical foot up to heights of 30 feet, exclusive of the cost of the timber. Towers higher than 50 feet are usually built of sawn timber bolted together, and one 150 feet high makes a heavy item in the expense of triangulation. At some stations no towers are required as the instrument may be directly upon the ground. Even in such cases a low tower ten feet in height is to be recommended, as it adds much to the comfort of the observer in warm weather, and has the advantage of elevating the line of sight above the surrounding earth.

The four posts of the exterior scaffold should be extended

about eight feet above the platform so as to allow canvas to be spread to protect the instrument from the sun and wind. The effect of the sun on the interior tripod is in high towers often very marked, the top moving in a lateral direction so as to describe an ellipse. To lessen this effect, and also for protection against wind, it is often screened by a canvas covering placed around the upper part of the scaffold.

The views of two triangulation stations of the U. S. Coast and Geodetic Survey here given may be of interest to students. The first shows a tower 160 feet in height erected by Mosman at Tate, Ohio, and the second the method used by the author at Port Clinton, Pa., where no tower was required. In the

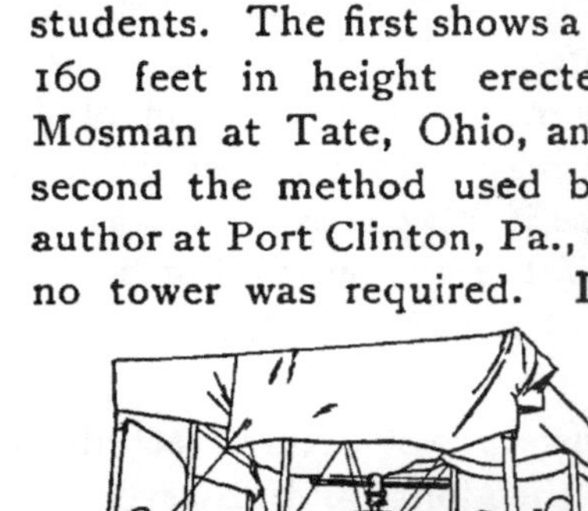

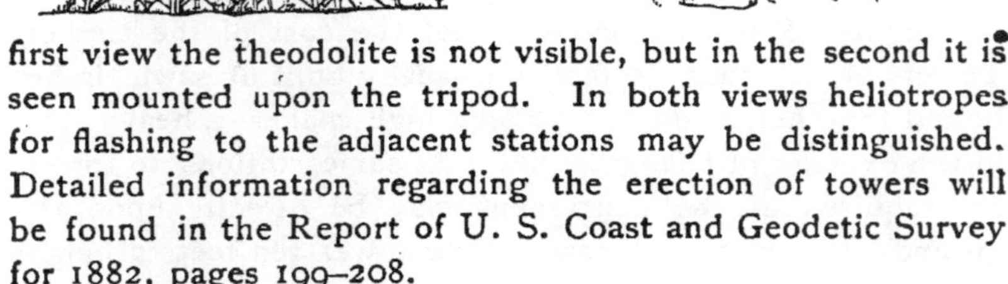

first view the theodolite is not visible, but in the second it is seen mounted upon the tripod. In both views heliotropes for flashing to the adjacent stations may be distinguished. Detailed information regarding the erection of towers will be found in the Report of U. S. Coast and Geodetic Survey for 1882, pages 199–208.

Sometimes a church spire, or other inaccessible point, is used as a station and angles are measured at other stations by sighting upon it. This is of frequent occurrence in secondary triangulation, but should be avoided in primary work. Sometimes in primary work an eccentric station is occupied near the true one, the angles observed there, and their values then reduced to the true station. Let A be the true station and a the eccentric one, and let it be required to find the true angle MAN from the observed angle MaN. To do this the distance Aa must be carefully measured and also the angle AaM, and the distances AM and AN must be found from the triangulation. Let $Aa = d$, $AaM = \theta$, $AM = m$, $AN = n$, $MaN = a$, and $MAN = A$. Then, as the opposite angles made by the crossing lines are equal, $A + M$ equals $a + N$, and accordingly the required angle is

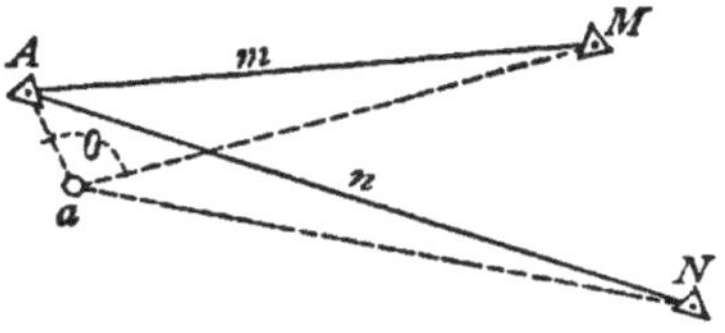

$$A = a - M + N, \qquad (72)$$

in which M and N are to be computed from

$$\sin M = \frac{d}{m}\sin\theta, \qquad \sin N = \frac{d}{n}\sin(\theta + a), \qquad (72)'$$

or, for most primary work, since m and n are large,

$$M = 206\,265\frac{d}{m}\sin\theta, \qquad N = 206\,265\,\frac{d}{n}\sin(\theta + a), \qquad (72)''$$

where M and N will be found directly in seconds. For example, let $d = 2.2145$ meters, $\log m = 3.90891$, $\log n = 3.95713$, $\theta = 28°\ 07'$, $a = 64°\ 18'\ 20''.13$. Then $M = 26''.550$ and $N = 50''.372$, whence $A = 64°\ 18'\ 43''.95$.

The angle θ is here measured from the fixed line aA around to the left-hand station and its value may range from 0° to 360°; hence the signs of M and N will depend upon the signs of $\sin\theta$ and $\sin(\theta + a)$. Thus if M and N were located at the left of a in the figure, θ would be over 180° and $\sin\theta$ would be negative. With regard to the use of (72)′ and (72)″ it may be said that the former need not be employed unless M and N are greater than 15 minutes.

Prob. 72. Draw the figure for the case where $d = 2.2145$ meters, $\log m = 2.90891$, $\log n = 2.95713$, $\theta = 208° \; 07'$, $a = 96° \; 07' \; 03''.72$, and compute the true angle A.

73. SIGNALS.

A signal is a pole, target, or other object erected at a station upon which the observer at another station points in measuring the angles. The simplest signal is a pole, but its use involves a liability to error in sighting upon the illuminated side, and hence for the most accurate work plane targets are preferred. These are made of a wooden framework covered with either black or white muslin. For a distance of fifteen or twenty miles good dimensions for a target are 2 feet in width and 12 feet in height. The target has the disadvantage of requiring to be set anew whenever the observer changes his station, but it has the advantage of being more easily seen than a pole. The old practice of putting a tin cone on a pole and of sighting on the illuminated side cannot be recommended, except for reconnaissance work.

For long lines neither pole nor target can be recognized, and the heliotrope must be used. This instrument consists essentially of a mirror which reflects the sunlight to the observer's station. The usual size of the mirror is about two inches in diameter, and it should be mounted so that it has a motion about a vertical and a horizontal axis. The mirror may be placed at one end of a board about three feet long upon which are two sights in the same line with the center of the mirror. The sights being pointed at the distant station, the mirror is constantly turned by an attendant, called a

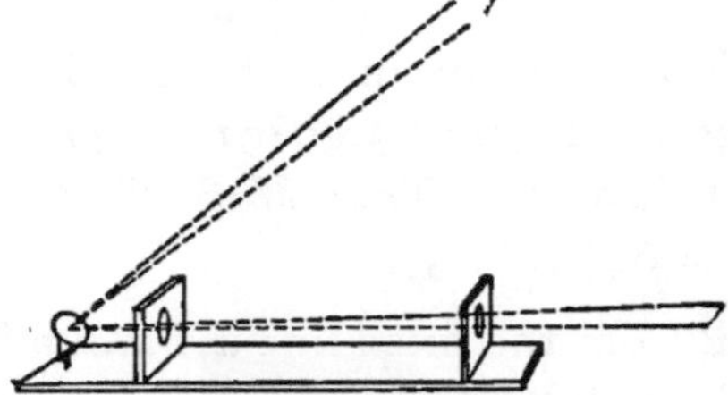

heliotroper, so that the shadow of the rear sight falls upon the front one, and the sunlight then is reflected to the observer, who sees it as a star twinkling in the horizon. As the apparent diameter of the sun is about half a degree, the reflected rays form a cone having the same angle, so that it is only necessary to point the heliotrope within a quarter of a degree of an object in order that the light may reach it. The light of a heliotrope may be seen through haze of moderate intensity if the observer knows where to point his telescope in order to find it.

Lines from ten to fifteen miles in length are usually observed with pole or target signals. For lines from fifteen to forty miles a combination of target and heliotrope is advantageous, the former being used on cloudy days and the latter in sunshine; in this case the heliotroper erects the target over the station and places his instrument in line in front of it. For lines exceeding fifty miles in length the heliotrope is the only feasible signal unless the atmosphere be unusually clear. Probably the longest side yet observed is one of 192 miles in California, where the heliotrope had a mirror of 77 square inches.

Night signals have been successfully used. These are generally large kerosene lamps with reflectors, which are placed in position and lighted by the heliotropers on leaving their stations in the evening. A magnesium tape whose burning is regulated by clockwork has been also employed. Night work should be usually combined with day work, the observer being on duty from noon to midnight. The best time for measuring horizontal angles is from six until nine o'clock in the morning, and from three in the afternoon until after sunset, as then the air is the clearest and the lateral refraction disturbances are the smallest. For vertical angles, on the other hand, the best time is during the two hours preceding and following noon, the vertical refraction being then the least variable.

In measuring horizontal angles it is sometimes necessary that a signal should be set at a short distance to one side of the center of the station. This is called the case of an eccentric signal, and a correction is to be applied to the observed angle to reduce it to the true angle. For instance, in 1878 an observer at the station O measured the angles COa and aOB, where the heliotrope had been set at a instead of at the true station A. The distance Aa was reported as 16 feet 2 inches, and the angle AaO as 129° 35′. Later, in 1883, the work had progressed so that OA was found to be 29 556 meters. The value of the small angle AOa in seconds is computed from $206\,265 Aa \sin AaO / AO$ and will be found to be 26″.63, and this is the correction to be added to COa and to be subtracted from aOB.

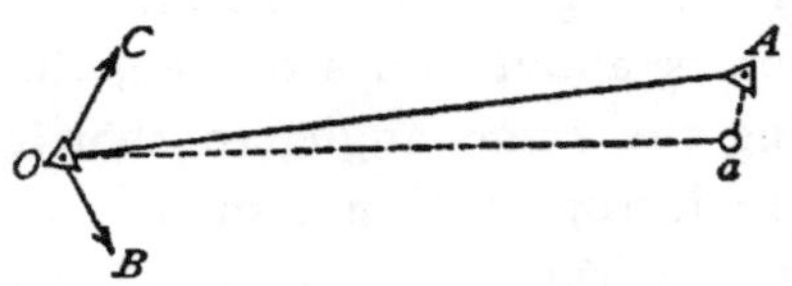

An eccentric signal should be avoided. Indeed it is best that heliotropers should not know that it can be used, otherwise they will be often tempted to set their heliotropes eccentrically from considerations of personal comfort and may neglect to take the measurements that are necessary for correcting the angles.

Prob. 73. Compute the correction AOa when the side OA is not very large compared with Aa, say when $OA = 295.56$ meters.

74. Horizontal Angles.

Two classes of triangulation are always recognized in geodetic work; the primary series, which connects directly with the bases and has the longest possible lines, and the secondary series, which locates stations within the primary triangles. To these are ultimately added a tertiary series for establishing stations at closer intervals for the special use of plane-table and stadia parties. It is generally required that the probable error of an observed value of a horizontal angle shall not exceed 0″.30 on primary work and 0″.80 on secondary work. On primary work repeating and direction theodolites are used, on secondary work repeating theodolites, while for the tertiary work the engineers' transit gives all the precision desirable. In fact a good engineer's transit will give as precise results as those required for secondary triangulation, provided the length of the lines be such that the signals can be clearly seen with its telescope.

A repeating theodolite does not differ in principle from an engineers' transit. The telescope, however, is so long that it cannot be turned over on its axis, but must be lifted out of the standards in order to be reversed in position. The graduated limb is usually from 8 to 12 inches in diameter, is divided into ten-minute divisions, and reads by three verniers to 3″ or 5″. Circles 16 and 20 inches in diameter were formerly used, but it is now known that the precision of these is little if any superior to those of 8 and 10 inches. The method of observation, in order to eliminate systematic and accidental errors, is in all respects the same as that described in Art. 14. Owing to the atmospheric disturbances on long lines of sight it is important that the work on each angle should be distributed over several days, and this is easy to arrange, since the rarity of good weather usually requires a party to remain two or three weeks at a station when several lines concentrate there.

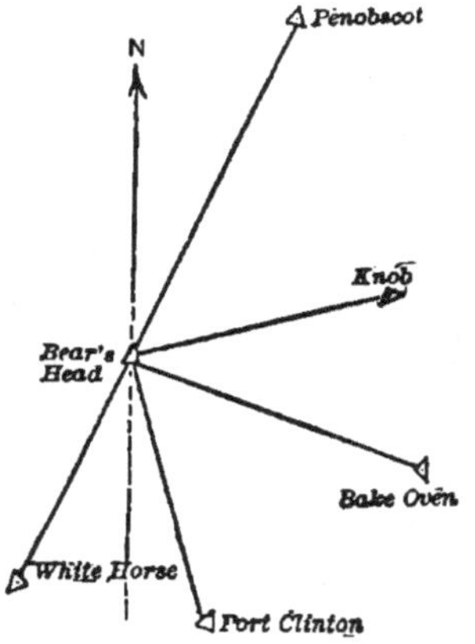

The following is a record of the work done with a repeating theodolite at Bear's Head station in Pennsylvania from July 20 to July 30, 1885. During these eleven days there were only eight when the weather permitted observations, and on five of

ANGLES AT BEAR'S HEAD.

Name of Angle.	No. of Reps.	Observed Value.	Adjusted.
Penobscot—Knob	48	51° 19′ 59″.71	60″.15
Penobscot—Bake Oven	40	85 52 32 .39	33 .22
Penobscot—Port Clinton	48	139 24 06 .97	06 .91
Penobscot—White Horse	40	180 39 48 .43	47 .24
Knob—Bake Oven	48	34 32 33 .16	33 .07
Knob—Port Clinton	48	88 04 05 .75	06 .76
Knob—White Horse	48	129 19 47 .57	47 .09
Bake Oven—Port Clinton	48	53 31 34 .27	33 .69
Bake Oven—White Horse	40	94 47 12 .71	14 .02
Port Clinton—White Horse	48	41 15 39 .97	40 .33

these no measurements could be made until about three o'clock in the afternoon. The total number of measures is seen to be 456, or an average of 114 for each independent angle. The station adjustment being made by the method of Art. 16, the average probable error of a single observed value is found to be 0″.71 and that of an adjusted value about 0″.60. It is thus seen that the adjustment has greatly increased the precision.

A direction theodolite has no verniers, but is read by three or more micrometer microscopes placed around the limb.

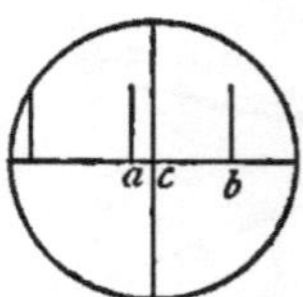

The circle in the figure represents the field of view of one of the microscopes in which three divisions of the graduated limb are seen. By turning the micrometer screw the cross-hair is moved to *a* or *b*, thus reading the distance *ac* or *bc* in seconds. When pointing on the first station the cross-hair may be set at a graduation mark, and when pointing at the second the reading is taken as just described. Such theodolites have large circles so that the limb may be divided to 5 minutes while the micrometers will read to seconds, and by taking the mean of all the micrometer readings a close determination of the angle can be made. No repetitions are possible by this method, but different series of readings are taken on different parts of the limb in order to eliminate errors of graduation, measures are made both from left to right and from right to left in order to eliminate errors due to clamping and twist, and the work is distributed over different days to eliminate atmospheric influences.

There are two methods of measuring the angles at a station with a direction theodolite. The first, called the method of single angles, is to determine each angle independently by the process above described; thus in the case of four lines meeting at O each angle is measured by reading first on the left-hand line and second on the right-hand line; thus the value found for BOC or BOD is independent of any reading

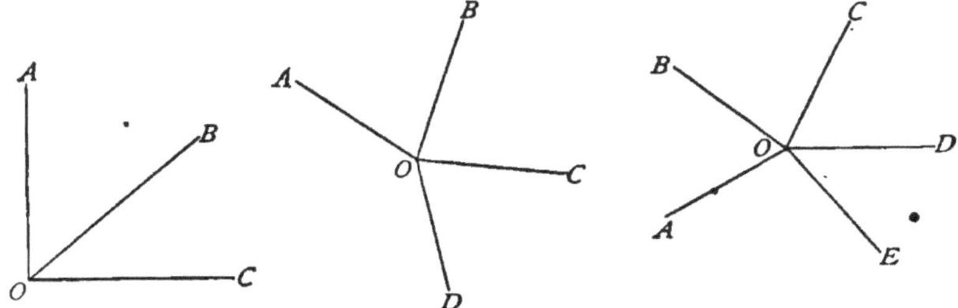

made on AOB. In this method all the results are to be treated and adjusted exactly as if they had been made by a repeating theodolite.

In the second method of observation, called the method of directions, a line OA is taken as a reference line and pointing and reading taken on it; then the limb is turned and readings taken on B, C, and D in succession. Another line OB is then taken as a reference line, and readings taken on C, D, and A in succession. Here it is seen that the values found are not independent, as the initial reading enters into all the results of each series; consequently the adjustment is more complicated than that of the other method.

Prob. 74. Regarding the above observations at Bear's Head station as of equal weight, compute the probable error of a single observed value.

75. The Station Adjustment.

The station adjustment for all cases except the method of directions is made by the method of Art. 16, which need not be further explained here. When the weights are so nearly equal as those of the case given in the last Article, it is an unwarrantable refinement to take them into account. With regard to the probable errors it is to be noted that those of the adjusted values need rarely be computed by the method of Art. 10 except in special scientific investigations. It is well, however, to find the probable error of a single observation by formula (10), and this ought to have a reasonable agreement with the average probable error of the observed values as computed from (9)′.

Some observers prefer to measure the n angles included between the n lines meeting at a station instead of combining the lines to make $\frac{1}{2}n(n-1)$ angles as in the example of the last Article. This case is called "closing the horizon," and thus the conditional equation is introduced that the sum of the n single angles shall be 360 degrees. The adjustment may be made by the method of Art. 16, employing only $n - 1$ independent quantities, but the numerical work will usually be shorter by the method of Art. 21.

The method of directions requires a slightly different process of station adjustment. To explain it take the case where the three lines OA, OB, and OC meet at the station O, and let x and y be the most probable values of the angles AOB and AOC. Suppose that OM denotes the direction of the telescope when the mean reading of the three microscope microm-

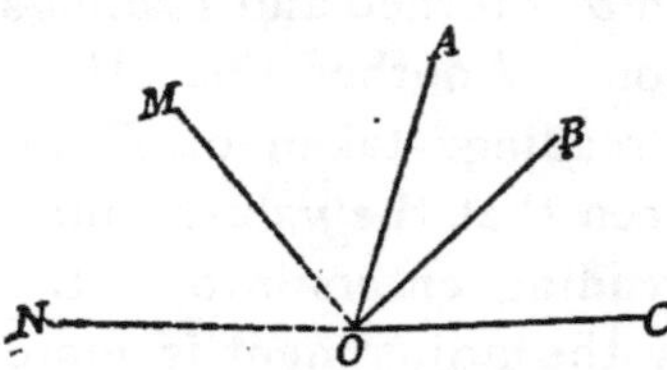

eters is 0° 00′ 00″.00, and let m denote the most probable value of MOA. Then let the three readings on A, B, and C give the three observation equations

$$\begin{aligned} m \qquad\quad &= 60^\circ\ 18'\ 20''.5, \\ m + x \qquad &= 85\ \ 04\ \ 13\ .0, \\ m \qquad + y &= 119\ \ 50\ \ 14\ .2. \end{aligned}$$

Next let the circle be turned so that ON gives the zero direction and let n be the angle NOA. Then three readings being taken on A, B, and C again, there are three additional observation equations

$$\begin{aligned} n \qquad\quad &= 120^\circ\ 17'\ 05''.0, \\ n + x \qquad &= 145\ \ 02\ \ 53\ .0, \\ n \qquad + y &= 179\ \ 48\ \ 59\ \ 5. \end{aligned}$$

Again if the circle be turned about 60 degrees further and three readings be taken upon A, B, and C there will be three more observation equations, while a fourth, fifth, and sixth set will each give three others. Thus for six positions of the circle there will be 18 observation equations involving 8 unknown quantities; from these the normal equations are formed by the rule of Art. 6, or, if they are of unequal weight, by the rule of Art. 7, and their solution will furnish the most probable values of x and y. The quantities m, n, . . . may be eliminated from the normal equations, before solving for x and y, as their numerical values are not required.

The numerical work may be abbreviated by introducing corrections to assumed values of the quantities. Thus, for the above case, let m_1 and n_1 be corrections to the observed values of m and n; also let $x = 24^\circ\ 45'\ 52''.5 + x_1$ and $y = 59^\circ\ 31'\ 53''.7 + y_1$. Then the six observation equations reduce to

$$\begin{aligned} m_1 &= 0, \qquad & m_1 + x_1 &= 0, \qquad & m_1 + y_1 &= 0, \\ n_1 &= 0, \qquad & n_1 + x_1 &= -\ 04''.5, \qquad & n_1 + y_1 &= +\ 0''.8, \end{aligned}$$

and from these the four normal equations are

$$
\begin{aligned}
3m_1 \qquad\qquad + x_1 + y_1 &= \quad 00''.0, \\
3n_1 + x_1 + y_1 &= -03.7, \\
m_1 + n_1 + 2x_1 \qquad &= -04.5, \\
m_1 + n_1 \qquad + 2y_1 &= +00.8.
\end{aligned}
$$

Taking the values of m_1 and m_1 from the first and second equations and substituting them in the others, these become

$$4x_1 - 2y_1 = -09''.8, \quad -2x_1 + 4y_1 = +06''.1,$$

from which $x_1 = -02''.25$ and $y_1 = +00''.40$ are the most probable corrections, whence $x = 24° 45' 50''.25$ and $y = 59° 31' 54''.10$ are the adjusted values of the angles AOB and AOC, and accordingly the most probable value of BOC is 24° 36′ 03″.85.

The angles found by the station adjustment are spherical angles, because the graduated circle is made level, that is parallel to a tangent plane to the spheroid at the station. Strictly speaking the level position of the graduated limb is an astronomical and not a geodetic one (Art. 59), but this slight discrepancy of a few seconds can produce no measurable effect on the observed angles. It should be borne in mind that it is of great importance to avoid inaccuracy of level when measuring angles, since this renders their values too large, and there is no method of eliminating its influence.

Prob. 75. Given the observed angles $AOB = 86° 07' 17''$ with weight 6, $BOC = 89° 10' 35''$ with weight 4, and $COA = 184° 41' 55''$ with weight 1. Compute the most probable values of the angles.

76. Triangle Computations.

After the angles have been measured at a number of stations and the length of one side has been obtained, either by connecting with an adjacent triangulation or by measuring it as a base, computations of the lengths of the triangle sides are to be made. The three angles of a triangle do not add up to 180 degrees and hence the results obtained for the sides are only approximate, but they are more than sufficiently accurate to compute the spherical excess of the triangle. These computations are the same in every respect as those explained in Art. 19, except that five-place logarithms should be used, the logarithmic sines taken to the nearest 10″ of angle, and the lengths determined only to the nearest 10 meters.

The formula for spherical excess established in (63)′ of Art. 63 may now be used and the excess be found for each triangle. In order to take the factor m from Table IV the mean latitude must be known roughly. In the first instance this may be estimated, but in later work it will be found from the results of the *LMZ* computations. Then,

$$\text{Spherical excess} = m \,.\, ab \sin C,$$

in which a and b are any two sides of the triangle and C is the angle included between them.

As a numerical example of the computation of spherical excess the following data of a triangle will be used:

Stations.	Angles adjusted at Stations.	Approximate Distances.	Approximate Latitudes.
Pimple Hill	49° 04′ 50″.13	27 540 meters	41° 02′
Smith's Gap	90 21 25 .53	36 440 meters	40 49
Bake Oven	40 33 46 .91	23 700 meters	40 45
	Sum = 180° 00′ 02″.57		Mean L. = 40° 52′

Now C can be taken as any one of these angles and a and b as the two adjacent sides. It is advisable to make two check computations for the excess, thus:

Numbers.	Logarithms.	Numbers.	Logarithms.
factor m	$\bar{9}.40441$	factor m	$\bar{9}.40441$
$a = 36\,440$	4.56158	$a = 23\,700$	4.37475
$b = 23\,700$	4.37475	$b = 27\,540$	4.43996
$C = 49°\ 04′\ 50″$	$\bar{1}.87831$	$C = 90°\ 21′\ 30″$	$\bar{1}.99999$
Excess = 01″.66	0.21905	Excess = 01″.66	0.21911

The adjustment of the angles of a spherical triangle is to

be made, when the angles are of equal weight, by applying to each of the given angles one-third of the discrepancy between the theoretic sum and the actual sum (Art. 18). For instance, using the above triangle, the correction to be

Stations.	Angles adjusted at Stations.	Spherical Angles.	Plane Angles.
Pimple Hill	49° 04′ 50″.13	49″.83	49″.28
Smith's Gap	90 21 25 .53	25 .22	24 .66
Bake Oven	40 33 46 .91	46 .61	46 .06
Sum =	180° 00′ 02″.57	01″.66	00″.00
180° + Excess =	180 00 01 .66		
Discrepancy =	− 00 .91		

subtracted from each given angle is 00″.30, and thus are found the adjusted spherical angles whose sum is 180° 00′ 01″.66. Then, to find the plane angles between the chords of the spherical arcs, one-third of 01″.66 is subtracted from each spherical angle.

When the weights of the given angles are very unequal it is advisable to take them into account by the method of (18). Thus if the triangle *KPS* have 8 sets measured on *K* and 48 on both *P* and *S*, and if the computed spherical excess is 01″.83, the spherical angles are found by applying corrections

Stations.	Weights.	Angles adjusted at Stations.	Spherical Angles.	Plane Angles.
K	1	41° 20′ 34″.34	35″.52	34″.91
P	6	79 03 41 .73	41 .93	41 .32
S	6	59 35 44 .18	44 .38	43 .77
	Sum =	180° 00′ 00″.25	01″.83	00″.00
	180° + Excess =	180 00 01 .83		
	Discrepancy =	+01 .58		

proportional to the weights, and then the plane angles found by diminishing each spherical angle by one third of the excess.

After these adjustments have been completed a second computation of triangle sides is to be made with seven-place logarithms and using, of course, the plane angles just found. The method is in all respects identical with that exemplified in Art. 19. The following form, which may be used for this computation, shows the angles at the stations, the adjusted spherical angles, and the plane angles, the spherical excess for this case being 02″.82. The triangle sides thus computed are the chords of the spherical arcs on the surface of the spheroid, the length of the base having been reduced to that surface by the method of Art. 31.

COMPUTATION OF A SPHERICAL TRIANGLE.

Lines and Stations.	Angles at Stations.	Corr.	Spherical Angles.	Sph. Excess.	Distances and Plane Angles.	Logarithms.
AB					43075 .54	4.6342308
C	54° 58′ 08″.84	+0″.08	08″.92	−0″.94	07″.98	0.0868007
A	95 29 01 .87	+0 .08	01 .96	−0 .94	01 .02	1̄.9980079
B	29 32 51 .86	+0 .08	51 .94	−0 .94	51 .00	1̄.6929746
CB					52364 .80	4.7190394
CA					25942 .16	4.4140061

The spherical angles here determined give the azimuths of *AC* and *BC* when the azimuths for the other side are known. Thus, if the azimuths of *AB* and *BA* are 204° 10′ 36″.05 and 24° 14′ 07″.92, the azimuth of *AC* is 299° 39′ 38″.01 and that of *BC* is 350° 41′ 15″.98. If the latitude and longitude of *A* and *B* are known, the two *LMZ* computations for finding the latitude and longitude of *C* and the azimuths of *CA* and *CB* may now be made by the method of Art. 66, the logarithms of the lengths of the sides being transferred from the above form. Then the next triangle, having *AC* or *BC* as its base, may be treated in like manner, and thus from one measured base and one astronomical station a chain of simple triangles is adjusted and computed.

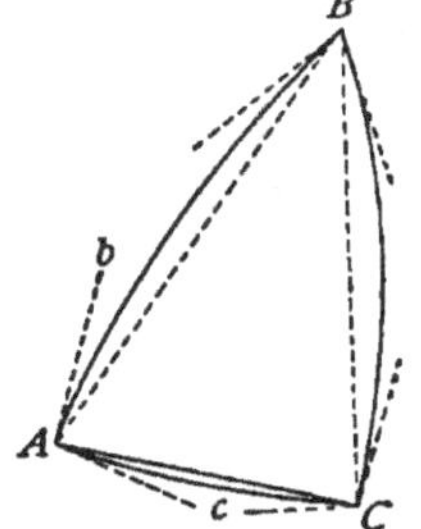

Prob. 76. Make all the computations described in this Article for the data of the following triangle, taking the angles as of equal

Stations.	Angles adjusted at Stations.	Approximate Latitudes.
Knob	50° 37′ 17″.20	40° 54′
Bake Oven	98 37 05 .05	40 45
Smith's Gap	30 45 41 .35	40 40

weight, the length of the side opposite Knob as 27535.63 meters, and the azimuth from Bake Oven to Smith's Gap as 252° 26′ 55″.42.

77. The Figure Adjustment.

There are two classes of conditions to be satisfied in the adjustment of a geodetic triangulation, those arising at the stations and those arising from the geometry of the figure.

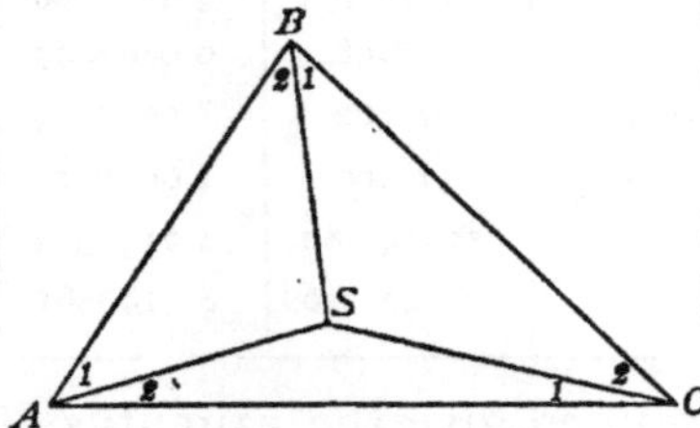

The station adjustment has already been discussed, and now the figure adjustment is to receive attention. This figure adjustment gives rise to conditions of two kinds, called angle conditions and side conditions. The requirement that the sum of the adjusted spherical angles of a triangle shall equal 180 degrees plus the spherical excess is an angle condition, while the requirement that the length of any side shall have the same value by whatever route it be computed is a side condition. For instance, in the figure *ABCS* there are four station adjustments to be made, if the angles are measured at the four stations; then, the figure adjustment requires that three angle conditions and one side condition shall be satisfied.

The strict method of making the adjustment of the case shown in the above figure is to state observation equations involving the angles at the stations, and conditional equations involving all the requirements of both station and figure adjustments. Thus, if three angles be measured at each station there will be twelve observation equations; for illustration suppose the angles have been measured only to the nearest degree, and that the observation equations are

$$A=84°,\ A_1=40°,\ A_2=43°,\quad B=56°,\ B_1=30°,\ B_2=27°,$$
$$C=40°,\ C_1=19°,\ C_2=21°,\quad S_1=106°,\ S_2=135°,\ S_3=120°,$$

where S_1 is the angle subtended by AB, and S_2 and S_3 those subtended by BC and CA. Then, the station adjustments give the four conditions

$$A=A_1+A_2,\quad B=B_1+B_2,\quad C=C_1+C_2,\quad S_1+S_2+S_3=360°.$$

The figure adjustment, supposing the triangles to be plane ones, requires the three angle conditions

$$A_2+B_1+S_1=180°,\quad B_2+C_1+S_2=180°,\quad C_2+A_1+S_3=180°,$$

and also, as shown in Art. 22, the side condition

$$\sin A_1 \sin B_1 \sin C_1 = \sin A_2 \sin B_2 \sin C_2.$$

The problem now is to determine the most probable values of the twelve observed angles which at the same time satisfy the eight conditional equations.

This problem is capable of rigorous solution, but when a figure contains many triangles it leads to very laborious computations. The custom has hence arisen of dividing the work into two parts; first, the station adjustments are made, each independently of the others, and secondly the values found by these station adjustments are then corrected so as to satisfy all the conditions of the figure adjustment. The station adjustments are generally made in the field, but the figure adjustment, which is far more lengthy, is reserved for the office, and is made by the method of correlates that is explained in the next Article.

In applying these principles to a triangle net consisting of a chain of simple triangles, having one side AB measured as a base and all angles observed, it is seen that the figure adjustment has as many angle equations as there are triangles, and no side equations.

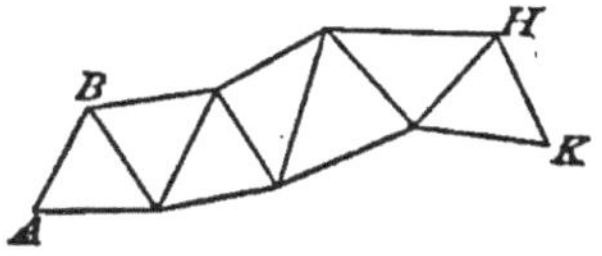

The figure adjustment is hence very simple, each triangle being treated in succession by the method of Art. 76, and the spherical angles and plane angles thus found are the final adjusted values. If, however, another side HK be also measured as a base, then a conditional side equation is to be introduced to express the requirement that the length HK as computed from AB shall be the same as the measured length; an illustration of this case for two triangles is given in the last paragraph of Art. 22.

In primary geodetic triangulation all stations are occupied and all lines sighted over in both directions. In secondary work a few of the stations may not be occupied, these being church spires or other inaccessible points. Thus in the last figure if one of the stations between B and H be not occupied the number of angle equations will be diminished by three, because there will be three triangles, in each of which one angle has not been observed and hence its value is to be found from those of the observed angles.

In stating the conditional equations that enter into a figure adjustment care should be taken to introduce no unnecessary ones, and the following rules will be useful for that purpose; these rules suppose only one base to have been measured. Let n be the total number of lines and n' the number of lines sighted over in both directions, let s be the total number of stations and s' the number of stations occupied for angle measurements. Then, in the figure adjustment,

$$\begin{aligned} &\text{Number of angle equations} = n' - s' + 1, \\ &\text{Number of side equations} \quad\; = n - 2s + 3. \end{aligned} \qquad (77)$$

For instance, in the figure $ABCS$, at the beginning of this Article, $n' = n = 6$, $s' = s = 4$, and hence there are three angle equations and one side equation; if, however, the station S had not been occupied, then $n' = 3$, $n = 6$, $s' = 3$, $s = 4$, and accordingly there would be one angle equation and one side equation.

Prob. 77. How many angle and side equations are there in the figure adjustment of each of the triangle nets shown in Art. 24, one base and the angles being measured?

78. Conditioned Observations.

By the proper selection of the unknown quantities it is generally possible to state observation equations so that these quantities will be independent (Art. 25), but a shorter method of adjustment, known as the method of correlates, may be established. In this method each observed quantity is represented by a letter and all the conditional equations are written, as in the illustration of the last Article. Let x, y, z, etc., represent the quantities whose values are to be found, and let the conditional equations be

$$a_1x + a_2y + \ldots = a,$$
$$b_1x + b_2y + \ldots = b,$$
$$c_1x + c_2y + \ldots = c,$$
$$\cdot \quad \cdot \quad \cdot \quad \cdot \quad \cdot \quad \cdot \quad \cdot \quad \cdot$$

in which the coefficients and constant terms are theoretic numbers. Now let M_1, M_2, M_3, ... be the values found by the observations for x, y, z, ...; if these values be inserted in the conditional equations they will not reduce to zero, owing to the errors of the measurements. Hence, let v_1, v_2, v_3, ... be small corrections which when applied to M_1, M_2, M_3, ... will render them the most probable values. Then if x, y, ... be replaced by $M_1 + v_1$, $M_2 + v_2$, ... the conditional equations reduce to

$$\begin{aligned} a_1v_1 + a_2v_2 + a_3v_3 + \ldots &= d_1, \\ b_1v_1 + b_2v_2 + b_3v_3 + \ldots &= d_2, \\ c_1v_1 + c_2v_2 + c_3v_3 + \ldots &= d_3, \\ \cdot \quad \cdot \quad \cdot \quad \cdot \quad \cdot \quad \cdot \quad \cdot & \end{aligned} \tag{78}$$

in which d_1, d_2, d_3, ... are small quantities called discrepancies. The problem now is to find values of v_1, v_2, v_3, ... which exactly satisfy these equations and which at the same time are the most probable values.

The following is the solution of this problem which is deduced in treatises on the Method of Least Squares. Let $p_1, p_2, p_3, \ldots$ be the weights of the observations M_1, M_2, $M_3, \ldots$ and let k_1, k_2, $k_3, \ldots$ be quantities which are determined by the solution of the normal equations

$$\left[\frac{a^2}{p}\right]k_1 + \left[\frac{ab}{p}\right]k_2 + \left[\frac{ac}{p}\right]k_3 + \ldots = d_1,$$
$$\left[\frac{ab}{p}\right]k_1 + \left[\frac{b^2}{p}\right]k_2 + \left[\frac{bc}{p}\right]k_3 + \ldots = d_2, \qquad (78)'$$
$$\left[\frac{ac}{p}\right]k_1 + \left[\frac{bc}{p}\right]k_2 + \left[\frac{c^2}{p}\right]k_3 + \ldots = d_3.$$
$$\cdot \quad \cdot \quad \cdot \quad \cdot \quad \cdot \quad \cdot \quad \cdot \quad \cdot \quad \cdot \quad \cdot \quad \cdot \quad \cdot \quad \cdot \quad \cdot$$

These equations are the same in number as the number of conditional equations, k_1 being known as the correlate of the first equation, k_2 of the second, and so on. The brackets indicate summation in accordance with the same notation as that employed in Art. 7, namely,

$$\left[\frac{a^2}{p}\right] = \frac{a_1^2}{p_1} + \frac{a_2^2}{p_2} + \ldots, \qquad \left[\frac{ab}{p}\right] = \frac{a_1b_1}{p_1} + \frac{a_2b_2}{p_2} + \ldots,$$

and the coefficients have similar properties to those in the normal equations for independent observations.

By the solution of these normal equations the values of k_1, k_2, $k_3, \ldots$ are found; then the corrections are

$$v_1 = \frac{a_1}{p_1}k_1 + \frac{b_1}{p_1}k_2 + \frac{c_1}{p_1}k_3 + \ldots,$$
$$v_2 = \frac{a_2}{p_2}k_1 + \frac{b_2}{p_2}k_2 + \frac{c_2}{p_2}k_3 + \ldots, \qquad (78)''$$
$$\cdot \quad \cdot \quad \cdot \quad \cdot \quad \cdot \quad \cdot \quad \cdot \quad \cdot \quad \cdot \quad \cdot \quad \cdot \quad \cdot$$

and these added to M_1, $M_2, \ldots$ give the most probable values of x, $y, \ldots$ which exactly satisfy the theoretic conditions. When there is but one conditional equation there is but one normal equation and one correlate, k_1, whose value is $d_1 \big/ \left[\frac{a^2}{p}\right]$, and thus the values of v_1, $v_2, \ldots$ agree with those deduced in Art. 21, where d_1 is called d, and q is used instead of a.

As an illustration of the method, let there be five measurements on five quantities, giving the observation equations,

$$\begin{array}{lll} 1. & x = 47.26, & \text{with weight } 3, \\ 2. & y = 39.04, & \text{with weight } 19, \\ 3. & z = 6.35, & \text{with weight } 13, \\ 4. & w = 86.64, & \text{with weight } 17, \\ 5. & u = 35.21, & \text{with weight } 6, \end{array}$$

which are subject to the two theoretical conditions,

$$x + y - w = 0, \qquad y + z - u = 10.$$

Let v_1, v_2, v_3, v_4, and v_5 be the most probable corrections to the observed values, so that the observation equations become

$$\begin{array}{lll} 1. & v_1 = 0, & p_1 = 3, \\ 2. & v_2 = 0, & p_2 = 19, \\ 3. & v_3 = 0, & p_3 = 13, \\ 4. & v_4 = 0, & p_4 = 17, \\ 5. & v_5 = 0, & p_5 = 6, \end{array}$$

and the conditional equations reduce to

$$\begin{array}{l} v_1 + v_2 \qquad - v_4 \qquad = + 0.34, \\ \qquad v_2 + v_3 \qquad - v_5 = - 0.18. \end{array}$$

Now, by comparison with the notation in (78),

$$\begin{array}{l} a_1 = +1,\ a_2 = +1,\ a_3 = \ \ 0,\ a_4 = -1,\ a_5 = \ \ 0,\ d_1 = +0.34, \\ b_1 = \ \ 0,\ b_2 = +1,\ b_3 = +1,\ b_4 = \ \ 0,\ b_5 = -1,\ d_2 = -0.18, \end{array}$$

and thus the normal equations of (78)′ become

$$\begin{array}{l} 0.445k_1 + 0.053k_2 = + 0.34, \\ 0.053k_1 + 0.296k_2 = - 0.18, \end{array}$$

whose solution gives $k_1 = + 0.855$ and $k_2 = - 0.759$. Then by (78)″ the values of the corrections, or residual errors, are

$$v_1 = +0.285,\quad v_2 = +0.005,\quad v_3 = -0.059,\quad v_4 = -0.050,\quad v_5 = +0.126,$$

and hence the adjusted values of the observations are

$$x = 47.545,\quad y = 39.045,\quad z = 6.291,\quad w = 86.590,\quad u = 35.336,$$

which are the most probable results that exactly satisfy the two conditional equations.

The probable error of an observation of the weight unity may be computed by the formula

$$r_1 = 0.6745\sqrt{\frac{\Sigma pv^2}{n - q + n'}},$$

in which n is the number of observation equations, q the number of unknown quantities, and n' the number of conditional equations. For the above example the residuals are already found; squaring them, multiplying each by its weight, and adding, gives $\Sigma pv^2 = 0.428$, whence $r_1 = 0.309$. Accordingly the probable error of the first observation is $0.309/\sqrt{3} = 0.18$, and the weight of the adjusted value of that observation must be somewhat smaller than 0.18.

Prob. 78. Four lines OA, OB, OC, and OD meet at a station O, and the following angles are observed, all of equal weight: $AOB = 19° 47' 13''$, $BOC = 40° 38' 04''$, $COD = 65° 12' 10''$, $DOA = 54° 22' 29''$, $BOD = 105° 50' 16''$, $DOA = 119° 30' 42''$. Let x, y, z, and w represent the four angles first named. Compute their adjusted values by the method of correlates.

79. Adjustment of a Polygon.

Let the diagram represent a polygonal figure having an interior station S, and let the angles which each side makes with the line to S be measured, S being an unoccupied station. By applying the rule of the last Article it is seen that there are two conditions in the figure adjustment, one angle equation and one side equation. If the figure be a plane one the angle condition is that the sum of the ten interior angles when adjusted shall be 540 degrees for a five-sided polygon. The side equation results from the

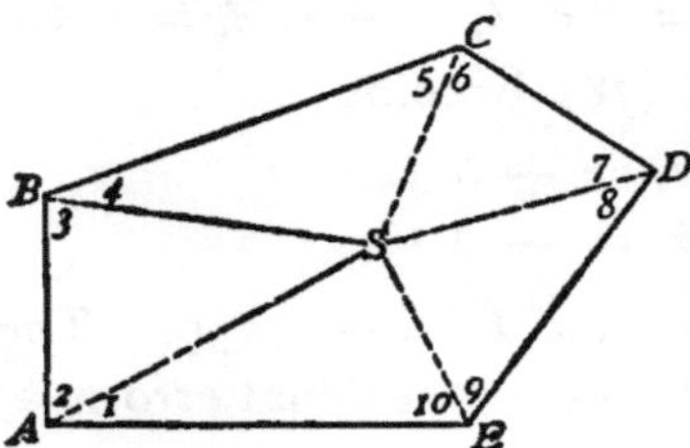

condition that, if one side be computed from another by two routes, the two expressions for its value shall be equal.

To express the first condition algebraically let v_1, v_2, . . . v_{10} be corrections in seconds to be added to the observed values of the angles. Let d_1 be the difference in seconds between the theoretic sum and the sum of the ten observed values; then

$$v_1 + v_2 + v_3 + v_4 + v_5 + v_6 + v_7 + v_8 + v_9 + v_{10} = d_1$$

is the conditional angle equation. To state the second condition let expressions for the side SD, as computed from SA by two routes, be written; if these be equated there results

$$\sin A_1 \sin B_3 \sin C_5 \sin D_7 \sin E_9 = \sin A_2 \sin B_4 \sin C_6 \sin D_8 \sin E_{10}$$

as the conditional side equation. This is to be expressed in terms of the corrections in a similar manner to that used in Art. 22, log $(A_1 + v_1)$ being written as log $A_1 + v_1$ diff. 1″, where diff. 1″ is the tabular logarithmic difference of the logarithmic sine corresponding to the angle A_1.

Let the observed values be those written below, all being of equal weight. As the sum of these is 540° 00′ 10″ the discrepancy d_1 is $-$ 10″, the angle equation is known, while

Observed Angles.	Log. Sines.
$A_1 = 25°\ 47'\ 23''$	$\bar{1}.6385588 + 43.5v_1$
$B_3 = 56\ \ 31\ \ 22$	$\bar{1}.9212208 + 14.0v_3$
$C_5 = 85\ \ 28\ \ 57$	$\bar{1}.9986487 + 1.7v_5$
$D_7 = 83\ \ 12\ \ 39$	$\bar{1}.9969439 + 2.5v_7$
$E_9 = 41\ \ 16\ \ 15$	$\bar{1}.8192933 + 23.9v_9$
	$\bar{1}.3746655$
$A_2 = 50°\ 12'\ 54''$	$\bar{1}.8856162 + 17.5v_2$
$B_4 = 48\ \ 52\ \ 12$	$\bar{1}.8769214 + 18.4v_4$
$C_6 = 61\ \ 58\ \ 02$	$\bar{1}.9458027 + 11.2v_6$
$D_8 = 38\ \ 25\ \ 07$	$\bar{1}.7933543 + 26.5v_8$
$E_{10} = 48\ \ 15\ \ 19$	$\bar{1}.8728079 + 18.8v_{10}$
	$\bar{1}.3745025$

the side equation is found by equating the two sums of the logarithmic sines. Thus

$$v_1 + v_2 + v_3 + v_4 + v_5 + v_6 + v_7 + v_8 + v_9 + v_{10} = -10'',$$
$$43.5v_1 - 17.5v_2 + 14.0v_3 - 18.4v_4 + 1.7v_5 - 11.2v_6 + 2.5v_7 - 26.5v_8 + 23.9v_9 - 18.8v_{10} = -1630,$$

where the second member of the last equation is in units of the seventh decimal place of logarithms.

By the method of the last Article the solution is now readily made, placing $a_1 = +1$, $a_2 = +1$, ... $d_1 = -10$ and $b_1 = +43.5$, $b_2 = -17.5$, ... $d_2 = +1630$. The two correlative normal equations are found to be

$$+10k_1 - 6.8k_2 = -10, \qquad -6.8k_1 + 4494.5k_2 = -1630,$$

from which $k_1 = -1.248$ and $k_2 = -0.364$. Then by (78)″

$$v_1 = -17''.1, \quad v_3 = -6''.3, \quad v_5 = -1''.9, \quad v_7 = -2''.2, \quad v_9 = -9''.9,$$
$$v_2 = +5''.1, \quad v_4 = +5''.5, \quad v_6 = +2''.8, \quad v_8 = +8''.4, \quad v_{10} = +5''.6,$$

are the most probable corrections to the observed values, and applying them the sum of the adjusted values will be found to be exactly 540°, and then the angles at S may be obtained. Also, multiplying each v by its tabular difference, the corrections to the logarithms may be found, and the sums of the two sets should then be exactly equal.

For a large polygon where the spherical excess of the triangles can be detected the method of adjustment is the same, the two conditional equations being slightly modified. First, the theoretic sum of the ten angles exceeds 540 degrees by two-thirds of the spherical excess of the entire polygon; thus if this excess be $18''.0$ the discrepancy d_1 will be $18''.0 - 10''.0 = +8''.0$. Secondly, each observed angle is to be diminished by one-third of the spherical excess of its triangle before placing it in the side equation; for instance, if the spherical excess of the triangle ABS is $03''.0$, then the value of A_1 to be used in the side equation is $25° 47' 22''$. The solution is now made as before and the corrections v_1, v_2, ... v_{10} found; these, added to the angles used in the side equations, give the adjusted plane angles, or, added to the observed values, they give the adjusted spherical angles.

When the station S is occupied and all the angles there are observed there will be five angle equations and one side equation in the figure adjustment. The side equation is the same as before and the five angle equations may be taken as those expressing the conditions that the sum of the angles in each triangle shall equal its theoretic value. Thus, for the triangle ABS, if the sum of the observed angles be 180° 00′ 05″ and the spherical excess be 03″.0, the angle equation is $v_2 + v_3 + v_{11} = +02''$. These six equations lead to six correlative normal equations, by whose solution the six correlatives are found, and then the fifteen corrections are obtained. Lastly, the adjusted spherical angles result by adding these corrections to the observed values, and the adjusted plane angles are found by subtracting from the spherical angles the proper amount for spherical excess. It may be remarked, however, that this solution can be abbreviated by an artifice similar to that used in the next Article.

Prob. 79. In the first diagram of Art. 77 let there be given $A_1 = 40\frac{1}{3}°$, $A_2 = 43\frac{1}{3}°$, $B_1 = 29\frac{2}{3}°$, $B_2 = 26\frac{2}{3}°$, $C_1 = 19°$, $C_2 = 21°$. Adjust these observations so that the results shall satisfy all the conditions of the figure adjustment.

80. ADJUSTMENT OF A QUADRILATERAL.

In the quadrilateral $ABCD$ let the two single angles at each corner be equally well measured. The rule of Art. 77 shows that the figure adjustment requires three angle equations and one side equation. The three angle equations may be written by taking any three of the triangles and imposing the conditions that in each the sum of the adjusted values shall equal the theoretic sum; the three triangles that have the point B in common will be chosen for this purpose. Let d_1, d_2, d_3 be the discrepancies for these triangles, d_1 being that for the triangle whose large

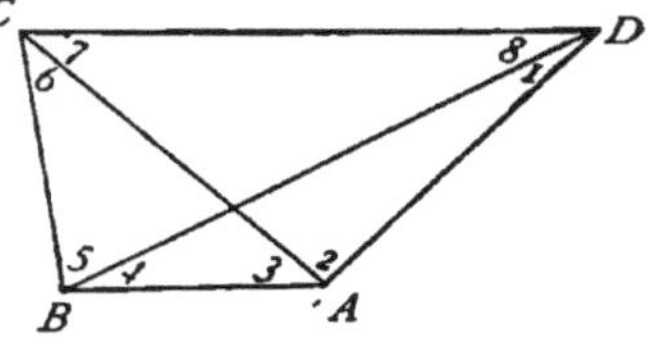

angle is A, while d_2 and d_3 are those for the triangles whose large angles are B and C; also let d_4 be the discrepancy for the fourth triangle CDA. Let $v_1, v_2, \ldots v_8$ be the corrections in seconds to be applied to the observed values. Then the three conditional angle equations are

$$\begin{aligned} v_1 + v_2 + v_3 + v_4 \qquad\qquad\qquad\qquad &= d_1 , \\ v_3 + v_4 + v_5 + v_6 \qquad\qquad &= d_2 , \\ v_5 + v_6 + v_7 + v_8 &= d_3 , \end{aligned}$$

and the conditional side equation is

$$\sin A_3 \sin B_5 \sin C_7 \sin D_1 = \sin A_2 \sin B_4 \sin C_6 \sin D_8 .$$

For given numerical values of the eight angles the adjustment may now be made by the method of Art. 78, there being four correlatives and four normal equations.

It is, however, frequently required to make an approximate adjustment, whereby the three angle equations will be satisfied, the side equation not being used. Taking, then, three correlatives, the normal equations (78)′ are, since all weights are unity,

$$\begin{aligned} 4k_1 + 2k_2 \qquad\quad &= d_1 , \\ 2k_1 + 4k_2 + 2k_3 &= d_2 , \\ 2k_2 + 4k_3 &= d_3 . \end{aligned}$$

Solving these and substituting the values in (78)″, the corrections are found, and, remembering that $d_2 + d_4 = d_1 + d_3$, these may be written

$$\begin{aligned} v_1 = v_2 &= \tfrac{1}{4}d_4 + \tfrac{1}{8}(d_1 - d_3), \\ v_3 = v_4 &= \tfrac{1}{4}d_2 + \tfrac{1}{8}(d_1 - d_3), \\ v_5 = v_6 &= \tfrac{1}{4}d_2 - \tfrac{1}{8}(d_1 - d_3), \\ v_7 = v_8 &= \tfrac{1}{4}d_4 - \tfrac{1}{8}(d_1 - d_3), \end{aligned} \qquad (80)$$

which are very easy in numerical application. For instance, let the three triangles have the spherical excesses $s_1 = 0''.48$, $s_2 = 1''.05$, $s_3 = 1''.41$, and $s_4 = 0''.84$, and let the observed values of the eight angles be arranged in four sets, one for each triangle. The sum of the observed angles for the first set subtracted from the theoretic sum gives the discrepancy

$d_1 = +02''.73$, and similarly for the other sets. Then by (80) the corrections are $+0''.62$, $+0''.74$, $+0''.67$, and $+0''.56$, whence result the adjusted values of the spherical angles. The sum of these is $360° \; 00' \; 01''.88$, which is a check on the work, since the spherical excess of the figure is $s_1 + s_3 = s_2 + s_4 = 1''.89$, the error of one unit in the second decimal being due to the lost digits in the third decimal. Lastly, the plane angles are found by subtracting the proper amounts from the spherical angles, these amounts being computed from (80) by using the given excesses instead of the discrepancies.

	Observed Angles.					Adjusted Angles.	
		A	*B*	*C*	*D*	Spherical.	Plane.
$D_1 =$	58° 44′	38″.98			38″.98	39″.60	39.51
$A_2 =$	25 18	16 .80			16 .80	17 .42	17.33
$A_3 =$	58 54	37 .54	57″.54			58 .28	58.14
$B_4 =$	37 02	04 .43	04 .43			05 .17	05.02
$B_5 =$	27 38		46 .48	46″.48		47 .15	46.77
$C_6 =$	56 24		09 .77	09 .77		10 .44	10.06
$C_7 =$	33 53			35 .14	35 .14	35 .70	35.38
$D_8 =$	62 03			27 .56	27 .56	28 .12	27.80
	179 59	57 .75	58 .22	58 .95	58 .48	01 .88	00.01
	180 00	00 .48	01 .05	01 .41	00 .84	01 .89	00.00
		+ 02 .73	+ 02 .83	+ 02 .46	+ 02 .36	00 .00	00.00

The adjusted values thus found will not, in general, satisfy the side equation, but by the following process a second series of corrections may be obtained that will insure this result. Let $v_1, v_2, \ldots v_8$ be the additional corrections to be applied to the above values of the plane angles. Then the angle equations are

$$v_1 + v_2 + v_3 + v_8 = 0,$$
$$v_3 + v_4 + v_4 + v_5 = 0,$$
$$v_1 + v_6 + v_7 + v_8 = 0,$$

and the side equation takes the form

$$a_1 v_1 + a_2 v_2 + a_3 v_3 + a_4 v_4 - a_5 v_5 - a_6 v_6 - a_7 v_7 - a_8 v_8 = d,$$

where $a_1, a_2, \ldots a_8$ are the tabular differences of the logarithmic sines corresponding to the values of the plane angles and d is the difference between the sums of the logarithmic sines of the even and odd angles. Now let

$$\begin{aligned} v_1 &= \quad u_1 + u_2, & u_4 &= \quad u_1 + u_4, \\ v_8 &= \quad u_1 - u_2, & u_5 &= \quad u_1 - u_4, \\ v_2 &= -u_1 + u_3, & v_6 &= -u_1 + u_5, \\ v_3 &= -u_1 - u_3, & v_7 &= -u_1 - u_5, \end{aligned}$$

and thus the angle equations are satisfied, while

$$(a_1 + a_4 + a_6 + a_7 - a_2 - a_3 - a_5 - a_8)u_1 + (a_1 + a_8)u_2 \\ + (a_2 - a_3)u_3 + (a_4 + a_5)u_4 + (-a_6 + a_7)u_5 = d$$

is the side equation in terms of the new quantities. This may be treated by the method of Art. 78, and after the u's are found, the values of v are known. For the above numerical case this side equation becomes

$$59.9u + 57.4u_2 - 15.2u_3 + 54.2u_4 - 20.1u_5 = +3,$$

where the second member is in units of the seventh decimal place of the logarithms. Then the single correlative equation gives $k_1 = 0.000288$, whence $u_1 = +0''.0017$, and the other u's are smaller still; accordingly the corrections v_1, v_2, . . . v_8 do not in any case amount to one one-hundredth of a second. The final adjusted values of the angles are hence those above given; had the corrections due to the side equations been appreciable they would have been added to both spherical and plane angles in order to give the final adjusted results.

Prob. 80. All the angles at stations A, B, and C are measured, but none at D. Find the number of equations in the figure adjustment and state them.

81. Final Considerations.

The preceding principles will enable the student to adjust and compute any common triangle net having but one measured base. If the net be composed of triangles only, each succeeding the other, the adjustment is made by starting with the base and computing each triangle in succession by the method of Art. 76. If it be composed of polygons only, each is separately adjusted by Art. 79, and the triangle sides then computed from the plane angles. If it be composed of

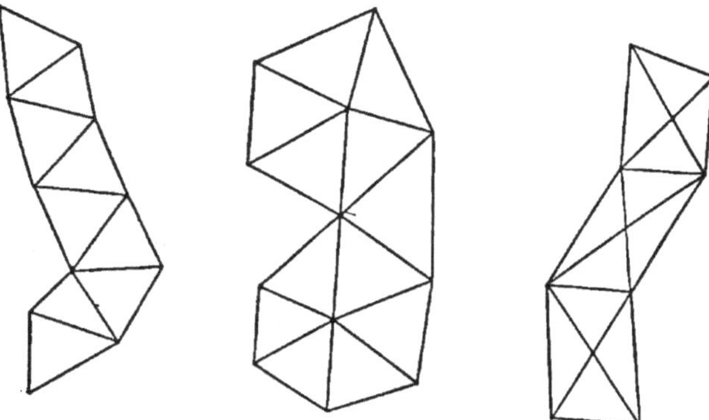

quadrilaterals only, the adjustment of each is made by Art. 80, and the sides then found from the plane angles. If it be made up of triangles, polygons, and quadrilaterals, as is generally the case, the same process may be followed, by starting at the base and treating each part in succession. Lastly the latitudes, longitudes, and azimuths are computed by the method of Art. 66 if the triangulation be a geodetic one, or by that of Art. 19 if it be plane.

When two bases are measured, or two stations occupied for astronomical work, the adjustment becomes so complex that it cannot be discussed in this elementary book. Such adjustments can only be successfully done by an office force specially trained in precise computation, and several weeks are perhaps required to solve the correlate normal equations that arise; this work is to be carried out in a systematic manner so that constant checks on the accuracy of the numerical work may be secured, and the probable errors of the final adjusted values may be determined.

It is not difficult to measure a single angle so that the final mean shall have a probable error as low as 0″.5, but the probable error of the adjusted value of the same angle as found from the figure adjustment will usually be somewhat larger. In this case the comparison of the probable errors does not perhaps give the fully correct idea, for there can be no doubt but that the figure adjustment has been most useful in eliminating accidental errors due to the pointings on signals, which in the measurement of a single angle may perhaps be a constant source of error. Undoubtedly the precision of the results of the figure adjustment is materially greater than that of the angles as determined by the station adjustment.

Throughout the entire field and office work all the coordinates, distances, and azimuths are to be regarded as approximate until the final figure adjustment is finished and the *LMZ* computations based on these are completed. In the beginning of the field work the latitudes and longitudes are known very roughly, being taken from such maps as are available, or found from compass readings and estimated distances. Later they become known to within a minute from rough triangle and *LMZ* computations, and at the close of the season's field work the various determinations of a station should check within a tenth of a second. After the office work of adjustment is finished, however, the latitudes and longitudes will agree to thousandths of a second, and then the triangle net may be regarded as definitely completed.

During the progress of the field work vertical angles are often taken by the method briefly described in Chapter IV. Such angles are to be measured by an instrument having a full vertical circle so that the double altitude, or double zenith distance, may be obtained by reversal. This vertical angle work, although its results cannot compare in precision with that of spirit leveling, furnishes valuable information in a new country which will repay its slight cost.

Boundary lines between countries are run most accurately after a triangulation has covered a strip along the general route. The location of the stations being thus determined, that of the boundary line is computed from the principles of geodesy and then points are set upon it by running out traverses from the stations. Many boundary lines have been run by determining astronomically the latitudes and longitudes of stations and then running out traverses and deflection lines from these. Owing, however, to the uncertainty of the plumb-line deflections these boundaries cannot compare in precision with those determined from a geodetic triangulation.

Prob. 81. Consult Report of the Commission on the Survey of the Northern Boundary of the United States in (Washgton, 1878,) and explain how points were located on the 49th parallel of north latitude.

Chapter X.

THE FIGURE OF THE EARTH.

82. The Earth as a Spheroid.

In the preceding chapters the fundamental principles for determining the size and eccentricity of the spheroid that best represents the earth have been presented, and the methods for computing geodetic surveys for given spheroidal elements have been explained. Now in conclusion a few other methods will be briefly discussed by which the dimensions or oblateness of the spheroid may be determined.

Pendulum observations give information regarding the ellipticity of the spheroid, since the length of a pendulum beating seconds is proportional to the force of gravity and since this force is greater in the polar than in the equatorial regions. Clairaut in 1743 deduced a remarkable theorem for the length of the seconds' pendulum at any latitude, namely,

$$s = S + (\tfrac{5}{2}k - f)S \sin^2 L, \qquad (82)$$

in which s is the length at the latitude L, and S is the length at the equator, k the ratio of the centrifugal force at the equator to the force of gravity, and f the ellipticity of the earth regarded as an oblate spheroid. This theorem is limited only by the assumptions that the earth is a spheroid rotating on its axis, and that its material is homogeneous in each of the concentric spheroidal strata. Now if the values of S and $(\frac{5}{2}k - f)S$ can be found from observations, then $\frac{5}{2}k - f$ is known, and since from the principles of mechanics k can be closely ascertained, the ellipticity f is determined.

For example, the following are a few of the many observations that have been made on the length of the seconds' pendulum:

Place.	Latitude.	Length of Pendulum.
Spitzbergen	79° 49′ 58″	39.2147 inches
Hammerfest	70 40 05	39.1952
London	51 31 08	39.1393
New York	40 42 43	39.1017
Jamaica	17 56 07	39.0351
Sierre Leone	8 29 28	39.0200
St. Thomas	0 24 41	39.0207

For each of these observations there may be written an observation equation of the above form; letting T represent the coefficient of $\sin^2 L$ the first one is

$$39.2147 = S + 0.96884T,$$

and similarly for each of the others. Then, applying the Method of Least Squares, the most probable values of S and T are found to be 39.0155 and 0.2021 inches. Accordingly the ratio of T to S is 0.005181, and this is the value of $\frac{5}{2}k - f$. But the value of k is about $\frac{1}{289}$ as found from the known facts regarding the intensity of gravity and the velocity of rotation at the equator; consequently the value of f is about $\frac{1}{288}$. Numerous discussions of pendulum observations appear to lead to the conclusion that the ellipticity of the earth, considered as a spheroid, is not far from $\frac{1}{288.5}$ or $\frac{1}{289}$. This is slightly larger than the value found from the discussions of meridian arcs, and the conclusion must hence be drawn that probably the spheroidal strata are not strictly homogeneous.

A theoretic discussion by Newton of the form assumed by a rotating homogeneous fluid under the action of gravity and centrifugal force led to the conclusion that the ellipticity was $\frac{1}{230}$. A similar one by Laplace indicates that f is about $\frac{1}{231}$. This value is, however, far too great, and it is accordingly indicated that the earth was not an homogeneous fluid at the time it assumed the present shape. For a full exposition of this branch of the subject reference is made to Todhunter's History of the Theories of Attraction and of the Figure of the Earth, London, 1873.

The shape of the earth may also be found from astronomical observations and computations. Irregularities in the motion of the moon were first explained by the deviation of the earth from a spherical form, and then these irregularities being precisely measured, the ellipticity may be computed, the value found by Airy being $\frac{1}{297}$, which is a little smaller than the result deduced from meridian arcs.

The size of the spheroid may also be deduced from measured arcs of a parallel between points whose longitudes are known. It is evident that such arcs have a special value in determining whether or not the equator and the parallels are really circles. The field work of a triangulation net extending across the American continent along the parallel of 39° north latitude was nearly completed in 1899 and the results of its discussion will soon be available. It may be noted, finally, that the elements of the spheroid may be deduced from a single geodesic line whose end latitudes and azimuths have been observed, or from such a line derived from a geodetic triangulation. The discussion of a geodesic line, extended through the Atlantic states from Maine to Georgia, by the U. S. Coast and Geodetic Survey, indicates that its influence upon our knowledge of the figure of the earth is to increase but slightly the dimensions of Clarke's spheroid of 1866 without appreciably changing his value of the ellipticity.

Three hundred and fifty years ago, when men first began to think about the shape of the earth on which it was their privilege to live, they called it a sphere, and they made rude measurements on its great surface to ascertain its size. These measurements, after nearly two centuries of work, reached an extent and precision sufficient to prove that its surface was not spherical. Then the earth was assumed to be a spheroid of revolution, and with the lapse of time the discrepancies in the data, when compared on that hypothesis, proved also that the assumption was incorrect. Granting that the earth is a sphere, there has been found the radius of one representing it more closely than any other sphere; granting that it is a spheroid, there has been also found, from the best existing data combined in the best manner, the dimensions of one that represent it more closely than any other spheroid. It has been seen that the radius of the mean sphere could only be found by first knowing the elliptical dimensions, and here it may be also thought that the best determination of the most probable spheroid would be facilitated by some knowledge of the theory of the size and shape of the earth considered under forms and laws more complex than those thus far discussed. In the following Articles, then, there will be given some account of the present state of scientific knowledge and opinion concerning the earth as an ellipsoid with three unequal axes, the earth as an ovaloid, and lastly the earth as a geoid.

83. THE EARTH AS AN ELLIPSOID.

As the sphere is a particular case of the spheroid, so the spheroid is a particular case of the ellipsoid. The sphere is determined by one dimension, its radius; the spheroid by two, its polar and equatorial diameters; while in the ellipsoid there are three unequal principal axes at right angles to each other that establish its form and size. As in the spheroid, the ellipsoid meridians are all ellipses, but the equator instead of being a circle is an ellipse of slight eccentricity. Let a_1 and a_2 denote the greatest and least semi-diameters of the equator of the ellipsoid, and b the semi-polar diameter; the ellipticities of the greatest and least meridian ellipses then are

$$f_1 = \frac{a_1 - b}{a} \quad \text{and} \quad f_2 = \frac{a_2 - b}{a_2},$$

while all other meridian ellipses have intermediate values. For the equator the ellipticity is $(a_1 - a_2)/a_1$. When the values of a_1, a_2, and b are known, the dimensions and proportions of the ellipsoid and of all its sections are fully determined.

On an ellipsoidal earth the curves of latitude, with the exception of the equator, are not plane curves, and hence cannot properly be called parallels. This results from the definition of latitude as may be seen from the diagram, where PP is the polar axis, $PQPQ$ the greatest meridian section, A a place of observation whose horizon is AA and latitude ABQ. Let now the least meridian ellipse, projected in the line PP, be conceived to revolve around PP until it coincides with the plane $PQPQ$ and becomes seen as $PQ'PQ'$. To find upon it a point A' that shall have the same latitude as A, it is only necessary to draw a tangent $A'H'$ parallel to AH touching the ellipse at A', then $A'B'$ perpendicular to $A'H'$ makes the same angle with the plane of the equator QQ as does AB. If the least meridian section be now revolved back to its true position, A' becomes projected at D'. Therefore, while a section through A parallel to the equator is an ellipse ADA, the curve joining the points having the same latitude as A is not plane, but a line of double curvature $AD'A$.

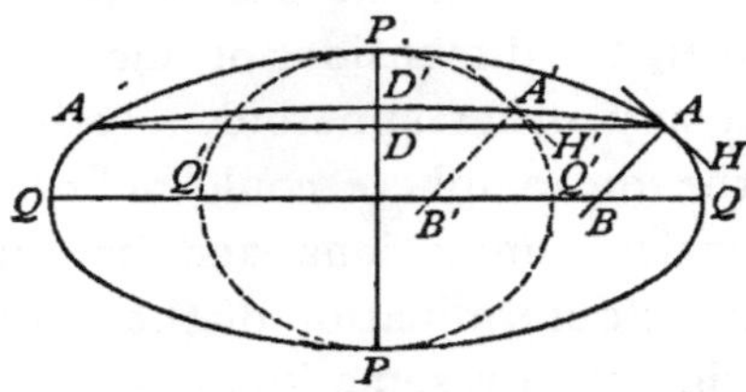

The process for determining from meridian arcs an ellipsoid to represent the figure of the earth does not differ in its fundamental idea from that explained in the last chapter for the spheroid. The normal to the ellipsoid at any point will usually differ slightly from the actual vertical as indicated by the plumb-line, and the sum of the squares of these deviations is to be made a minimum in order to find the most probable elements of the ellipsoid. An expression for the difference of these deviations at two stations on the same meridian arc is first deduced in terms of four unknown quantities, three being the semi-axes a_1, a_2, and b, or suitable functions of them, and the fourth the longitude of the greatest meridian ellipse, referred to a standard meridian such as that of Greenwich; and in terms of four known quantities, the observed linear distance between the two stations, their latitudes and the longitude of the arc itself. Selecting now one station in each meridian arc as a point of reference, there are written for that arc as many equations as there are latitude stations, inserting the numerical values of the observed quantities. These equations will contain four more unknown letters than there are meridian arcs, and from them as many normal equations are to be deduced as there are unknown quantities, and the solution of these will furnish the most probable values of the semi-axes a_1, a_2, and b, with the longitude of the extremity of a_1, and also the probable plumb-line deviations at the standard reference stations. The process is long and tedious, but it is easy to arrange a system and schedule, so that the computations may be accurately carried out and constant checks be furnished.

The first deduction of an ellipsoid to represent the figure of the earth was made in Russia, by Schubert, about the year 1859. He found $f_1 = \frac{1}{292}$ and $f_2 = \frac{1}{302}$ for the two meridian ellipticities, and $\frac{1}{8881}$ for that of the equator. The longitude of the ellipse of greatest eccentricity was found to be about 41° East of Greenwich, and the length of its quadrant was determined as 10 002 263 meters, that of the quadrant of least eccentricity being 10 001 707 meters.

It is, however, Clarke of the British Ordnance Survey to whom is due the credit of the most careful investigations in this direction. His discussion of 1866 included meridian arcs amounting in total to nearly five-sixths of a quadrant and containing 40 latitude stations, while his discussion of 1878 included the same data and several additional arcs. The following table gives the result of these discussions. The linear dimensions are in meters, the meter used being one equivalent to 3.28086933 feet and hence slightly longer than the U. S. legal meter that has been employed in the preceding Chapters.

CLARKE'S ELEMENTS OF THE ELLIPSOID.

	1866.	1878.
Greatest equatorial semi-diameter, a_1	6 378 294	6 378 380
Least equatorial semi-diameter, a_2	6 376 350	6 377 916
Polar semi-axis, b	6 356 068	6 356 397
Greatest meridian quadrant, q_1	10 001 553	10 001 867
Least meridian quadrant, q_2	10 000 024	10 001 507
Quadrant of the equator, Q	10 017 475	10 018 770
Greatest meridian ellipticity, f_1	1/287.0	1/289.5
Least meridian ellipticity, f	1/314.4	1/295.8
Ellipticity of the equator, F	1/3281	1/13706
Longitude of q_1	15° 34′ East	8° 15′ West

In comparing these ellipsoids the different positions of the greatest meridian attract notice; in the first it passes through Germany and in the second through Ireland, the angular distance between them being about 24 degrees. The least meridian, 90 degrees distant, passes through Pennsylvania for the first spheroid and through Kansas for the second. It would thus appear that the radius of curvature is greater on the American than on the European continent.

It seems to be the prevailing opinion that satisfactory elements of an ellipsoid to represent the earth cannot be obtained until geodetic surveys shall have furnished more and better data than are now available, and particularly data from arcs of longitude. The ellipticities of the meridians differ so slightly that measurements in their direction alone are insufficient to determine, with much precision, the form of the equator and parallels. In Europe and America several longitude arcs will soon be available, and it will then be possible to obtain more reliable elements of the spheroid. At present the ellipsoids represent the figure of the earth as a whole very little better than do the spheroids, although, for certain small portions, they may have a closer accordance. For instance, the average probable error of a plumb-line deviation from the normals to the Clarke ellipsoid of 1866 is $1''.35$, while for the spheroid derived from the same data it is $1''.42$. Further, the marked differences in the ellipticities of the equator of the two Clarke ellipsoids, due to comparatively slight differences in data, are not pleasant to observe. Lastly, the ellipsoid is a more inconvenient figure to use in calculations than the spheroid. For these reasons the earth has not yet been regarded as an ellipsoid in practical geodetic computations, and it is not probable that it will be for a long time to come.

84. The Earth as an Ovaloid.

In a spherical, spheroidal, or ellipsoidal earth the northern and southern hemispheres are symmetrical and equal; that is to say, a plane parallel to the equator, at any south latitude, cuts from the earth a figure exactly equal and similar to that made by such a plane at the same north latitude. The reasons for assuming this symmetry seem to have been three: first, a conviction that a homogeneous fluid globe, and hence perhaps the surface of the waters of the earth, must assume such a form under the action of centrifugal and centripetal forces; secondly, ignorance and doubt of any causes that would tend to make the hemispheres unequal; and thirdly, an inclination to adopt the simplest figure, so that the labor of investigation and calculation might be rendered as easy as possible. These reasons are all very good ones, but gradually there have arisen certain considerations leading to the conclusion that there are causes which tend to make the southern hemisphere greater than the northern. These considerations embrace a vast field of inquiry in astronomy and physical geography of which only a brief statement can be given here.

The earth moves each year in an ellipse, the sun being in one of the foci, and revolves each day about an axis inclined some 66½ degrees to the plane of that orbit. When this axis is perpendicular to a line drawn from the center of the sun to that of the earth occur the vernal and autumnal equinoxes, and at points equally removed from these are the summer and winter solstices. For many centuries the earth's orbit has been so situated in the ecliptic plane that the perihelion, or nearest point to the sun, has nearly coincided with the winter solstice of the northern hemisphere and the summer solstice of the southern hemisphere. The consequences are: first, the half of the year corresponding to the winter is about seven days longer in the southern hemisphere than in the northern; secondly, during the year the south pole has about 170 more hours of night than of day, while the north has about 170 more hours of day than of night; and, thirdly, winter in the northern hemisphere occurs when the sun is at his least distance from the earth, and in the southern when he is at his greatest. From these three reasons it would seem that the amounts of heat at present annually received by the two hemispheres should be unequal, the northern having the most and the southern the least. Now, on considering the physical geography of the globe, these two facts are seen: first, fully three-fourths of the land is in the northern hemisphere clustered about the north pole, while the waters are collected in the southern; and secondly, the south pole is enveloped and surrounded by ice to a far greater extent than the northern. There is then a considerable degree of probability that some connection exists between these astronomical and terrestrial phenomena, that the former, indeed, may be the cause of the latter. The mean annual temperature of the southern hemisphere may have been for many centuries sufficiently lower than that of the northern hemisphere to have caused an accumulation of ice and snow whose attraction drags the waters toward it, thus leaving dry the northern lands and drowning the southern ones with great oceans. Hence there appear to be causes which tend to render the earth ovaloidal, or egg-like, in shape, the large end being at the south pole.

The process of finding the dimensions of an ovaloid of revolution to represent the figure of the earth would be the same in principle as that already described for the spheroid and ellipsoid. The equation of an oval should be stated, and preferably one that reduces to an ellipse by the vanishing of a certain constant. From this equation an expression for the length of an arc of the meridian for both north and south latitude can be deduced, and this be finally expressed in terms of the small deviations between the plumb lines and the normals to the ovaloidal meridian section at the latitude stations. The solution of these equations by the Method of Least Squares will give the most probable values of the constants, determining the size and shape of the oval due to the data employed. Such computations have not yet been undertaken, on account of the lack of sufficient data from geodetic surveys in the southern hemisphere. Since such surveys can only be executed on the continents and largest islands, it is clear that the data will always be few in number compared with those from the northern hemisphere. Pendulum observations, discussed on the hypothesis of a spheroidal globe, by Clairaut's theorem, are able to give some information; since they can be made on small islands as well as on the main lands, it is possible thereby to obtain knowledge concerning the separate ellipticities of the two hemispheres.

An important idea to be noted in this branch of our subject is that the surface of the waters of the earth is, probably, not fixed, but variable. About the year 1250, the perihelion and the northern winter solstice coincided, and the excess in annual heat imparted to the northern hemisphere was near its maximum. Since that date they have been slowly separating, and are now nearly eleven degrees apart. This separation increases annually by about 61″.75, so that a motion of 180 degrees will require about 10 450 years, and when that is accomplished the perihelion will coincide with the southern winter solstice. Then the condition of things will be exactly reversed, and perhaps the ice will accumulate around the north pole, the waters will flow back from the south to the north, and the lands in the southern hemisphere become dry while those in the northern hemisphere become submerged. The change is so slow that it might remain undetected for centuries and yet ultimately be sufficient to perceptibly alter the relative shapes and sizes of the two hemispheres. These considerations, though interesting, are speculations only, for causes not now known may intervene to produce results which as yet have not been even faintly imagined.

85. The Earth as a Geoid.

The word Geoid is used to designate the actual figure of the surface of the waters of the earth. The sphere, the spheroid, the ellipsoid, the ovaloid, and many other geometrical figures may be, to a less or greater degree, sufficient practical approximations to the geoidal or earthlike shape, yet no such assumed form can be found to represent it with perfect accuracy. The geoid, then, is an irregular figure peculiar to our planet; so irregular, indeed, that some have irreverently likened it unto a potato; and yet a figure whose form may be said to be subject to fixed physical laws, if only the fundamental ideas implied in the name be first clearly and mathematically defined.

The first definition is that the surface of the geoid at any point is perpendicular to the direction of the force of gravity, as indicated by the plumb line at that point; from the laws of hydrostatics it is evident that the free surface of all waters in equilibrium must be parallel to that of the geoid. The second definition determines that our geoidal surface to be investigated is that coinciding with the surface of the great oceans, leaving out of consideration the effects of ebb and flood, currents and climate, wind and weather. Under the continents and islands this surface may be conceived to be produced so that it shall be at every place perpendicular to the plumb-line directions. If a tunnel be driven on the surface from ocean to ocean it is evident that the water flowing from each would attain equilibrium therein and its level would show the form of the geoid along that section of the earth.

The following figure may perhaps give a clearer idea of the properties of the geoid and of its relation to the spheroid. It represents a small part of a meridian section, the northern part

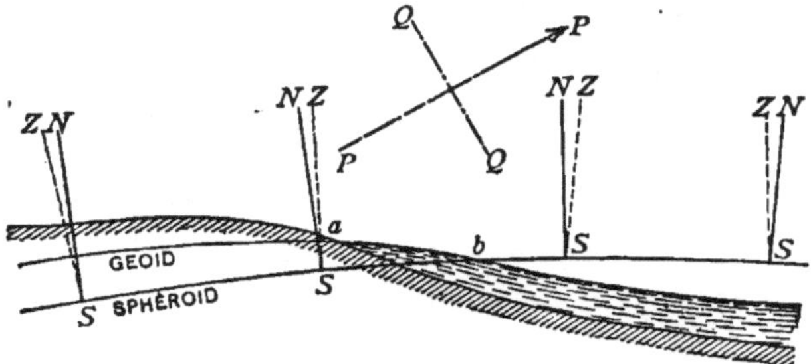

being land and the southern part being the ocean. The full-line curve shows the section of the spheroid, while the lighter line shows that of the geoid. At any station *S* the line *SN* is normal to the spheroid, while the line *SZ* is the direction of the plumb line or of the force of gravity. Hence the surface of the spheroid is normal at each station to the line *SZ*. The line *PP* being drawn parallel to the axis of the earth and *QQ* parallel to the equator, it is clear that the angle which *SZ* makes with *QQ* is the astronomical latitude of the station *S*, while the angle which *SN* makes with *QQ* is the geodetic latitude. The angle *ZNS* is hence the difference of these latitudes, or the so-called plumb-line deflection.

The figure represents roughly the probable relative positions of the spheroid and geoid. Under the continents the geoid tends to rise higher, while on the oceans it tends to sink lower, than the surface of a spheroid of equal volume. The attraction of the heavier and higher continents lifts, so to speak, the geoidal surface upward, while the lower and lighter ocean basins allow it to sink downward. To this rule there are, however, many exceptions, and these exceptions teach that the earth's crust is of variable density; for instance, southward of the great mountains in India it would be expected that the plumb-line deflections would all be toward the north, but this is by no means the case.

It may now be seen that the plumb-line deflections are really something artificial, depending upon the use of a particular spheroid. The geoid is an actual existing thing; the spheroid is not, but is largely an assumption introduced for practical and approximate purposes. At any station S in the above figure, the direction SZ is the only one that can be observed, and the angle made by it with QQ can be measured with a probable error of less than one-tenth of a second of arc. The angle ZSN, or the so-called plumb-line deflection at S, will hence vary with the elements of the particular spheroid employed, and with the correct orientation of geoid and spheroid. A geodetic latitude (or spheroidal latitude as it should perhaps be more properly called) is something that cannot be directly measured, and therefore it seems that the plumb-line deviations for even a particular spheroid cannot be absolutely found until observations have been made over an extent of country wide enough to enable us to judge of the laws governing the geoid itself. A very slight change in the position of the above elliptical arc may add or subtract a constant quantity from each of the angles between the true verticals and the normals. The differences of the plumb-line deflections at neighboring stations will, however, remain closely the same. For instance, if two plumb-line deflections are 3″.50 and 1″.25 for a certain spheroid, another spheroid may be drawn making them 2″.75 and 0″.50, the difference being 2″.25 in both cases.

The above figure gives a very exaggerated picture of the relation between spheroid and geoid. The greatest plumb-line deflections are about 30″, and it is unusual to find them exceeding 15″; this small angle could not, of course, be seen in a common drawing, and hence in any true representation the spheroidal and geoidal sections should be drawn parallel.

It is further to be noted that these plumb-line deflections in longitude as well as in latitude, and also that the astronomical and geodetic azimuths of a line, do not agree. Thus at Parkersburg station, on the U. S. Lake Survey, the plumb-line deflection in latitude was 1″.47 toward the south, the deflection in longitude was 0″.70 toward the west, and the discrepancy in azimuth was 0″.76. The greatest deflection in latitude found on this survey was 10″.77, and the greatest one in longitude 12″.13, while the greatest discrepancy in azimuth was 11″.56.

86. CONCLUSION.

There have now been briefly set forth a history of geodesy and the elements of geodetic theory and practice. It has been seen that the first idea of the shape of the earth was that of the plane, and the second that of the sphere. Assuming it to be a sphere measurements and computations were made to find its size; these being discordant it was assumed to be a spheroid and more elaborate investigations were undertaken. Assuming it to be an ellipsoid other computations were made. The question now arises whether the form of the geoid can be deduced so that definite statements can be made regarding its size and figure.

Compared with a spheroid of equal volume, the geoid has a very irregular surface, now rising above that of the spheroid, now falling below it, and ever changing the law of its curvature, so as to conform to the varying intensity and direction of the forces of gravity. Where the earth's crust is of greatest density there it rises, where the crust is of least density and thickness there it sinks. From a scientific point of view it will be valuable to know the laws governing its form and size; from a practical point of view it appears that until these are known the earth's figure can never be accurately represented by a sphere or spheroid or ellipsoid, or other geometrical form. For instance, if it be desired to represent the earth by an oblate spheroid, the best and most satisfactory one must be that having an equal volume with the geoid, and whose surface everywhere approaches as nearly as possible to the geoidal surface. Such a spheroid cannot, of course, be found until more and better data concerning the geoid have accumulated, yet what has already been said is sufficient to indicate that the dimensions at present used are probably somewhat too large. Granting that in general the geoid rises above this spheroid under the continents and falls below it on the seas it seems evident, since the area of the oceans is nearly three times that of the lands, that the intersection of the two surfaces will generally be some distance seaward from the coast line, as seen at *b* in the figure of the last Article. Now geodetic surveys can only be executed on the continents, and even if they be reduced to the sea level at the coast, or to *a* in the diagram, the elements of a spheroid deduced from them will be too large to satisfy the above condition of equality of volumes, for the ellipse through *a* is evidently larger than that through *b*. At present it would be purely a guess to state what quantity should be subtracted from the semi-axes of the Clarke spheroid on account of these considerations.

For a locality where precise astronomical and geodetic work has been done a fair picture of the relation of the spheroid to the geoid may be obtained. But on the oceans, where such work cannot be executed, it will generally be impossible to secure numerical comparisons. The word Geoid, in fact, with all the fruitful ideas therein implied, is comparatively new, it having been coined in 1872. Its mathematical properties, resulting from its definition, have been studied, and Bruns has demonstrated that the mathematical figure of the earth may be determined independently of any hypothetical assumption, provided that there have been observed at and between numerous stations five classes of data, namely, astronomical determinations of latitude, longitude, and azimuth, base line and triangulation measurements, vertical angles between stations, spirit leveling between stations, and determinations of the intensity of the forces of gravity. These five classes are sufficient for the solution of the problem, but also necessary; that is, if one of them does not exist, a hypothesis must be made concerning the shape of the earth's figure. These complete data have, however, never yet been observed for even an extent of country so small as England, a land probably more thoroughly surveyed than any other. To render geodetic results of the greatest scientific value, it is hence necessary that either the pendulum or some other instrument of precision should be employed to determine the relative intensity of the forces of gravity at the principal triangulation stations, and that trigonometric leveling by vertical angles should be brought to greater perfection.

These conclusions appear to neglect the circumstance that the geoid is not a fixed and invariable figure. Atmospheric influences are continually at work to tear down the mountain ranges and fill up the ocean basins; as this goes on the geoid tends to become more uniform in curvature. Internal fires cause parts of the earth's crust to slowly rise and fall, and immediately the geoidal surface undergoes a like alteration. These changes are, however, probably slight compared to those caused by the conical rotation of the earth's axis around its mean position in its period of about 425 days. Owing to this rotation all astronomical latitudes, longitudes, and azimuths are subject to periodic changes, and the position of the geoid with respect to the spheroid is constantly varying. It is hence plain that the geoid can never be used as a figure for standard reference in geodetic surveying. On the contrary, a spheroid of revolution, or an ellipsoid with three unequal axes, must always be employed as the standard figure, its size and shape being so determined as to render a minimum the sum of the squares of all the plumb-line deflections that occur during a complete cycle of the conical axial rotation.

CHAPTER XI.

TABLES.

87. EXPLANATION OF THE TABLES.

The following tables have been mostly compiled from the extended tabulations given in Reports of the U. S. Coast and Geodetic Survey and in the Smithsonian Geographical Tables. They will be sufficient for the solution of the problems given in this volume, although in some cases the larger tables will be more convenient in interpolation. All dimensions of the earth are for the Clarke spheroid of 1866.

The meter used in these tables agrees with that of the Smithsonian volume, its relation to the yard being exactly expressed by the fraction $\frac{3937}{3600}$ and to the foot by $\frac{3937}{1200}$; thus, the number of feet in a meter is 3.2808333; the meter of the tables of the Coast and Geodetic Survey published prior to 1897 is, however, 3.2808693 feet, this being the value deduced by Clarke's comparisons of 1866.

Table I contains mean values of the correction for refraction to be added to all vertical angles taken upon celestial objects. The values given are mean ones, the word mean implying average atmospheric conditions. Under unusual extremes of temperature and barometric pressure the actual refraction may be greater or less than these mean values, and in precise work thermometer and barometer readings are to be taken. For the class of work described in Chapter V the mean values of the table will be amply sufficient. Further, it may be remembered that for altitudes greater than 30° the mean refraction is expressed by $57''.7 \cot h$, where h is the apparent altitude.

Tables II and III contain lengths of arcs of the meridian and parallels in miles, meters, and feet. The length stated for a meridian arc is the length of an arc whose middle point has that latitude; thus the length of one degree of latitude at latitude 40° is the length from latitude 39° 30′ to latitude 40° 30′. The lengths given for arcs of longitude are measured on the parallels of latitude. The manner of computing these tables is explained in Art. 61.

Table IV contains logarithms of the radii of curvature of the meridian and of the prime-vertical normal section; the derivation of these values is explained in Art. 62. The addition of the two logarithms gives the logarithm of the square of the radius of the osculatory sphere, since $R_1R_2 = R^2$. The fourth column contains logarithms of the radii of curvature of the parallels, computed by the formula $r = R_2 \cos L$. The last column gives values of the logarithm of $206\,265/2R_1R_2$, which is to be added to the logarithm of twice an area, in meters, in order to obtain the logarithm of the spherical excess of the area, as explained in Art. 63.

Table V contains logarithms of the constants to be used in computing geodetic latitudes, longitudes, and azimuths. The expressions for these constants and the manner of obtaining their values are given in Art. 64. The factor E, given in the last column, will be needed in computing large triangles and is hence retained in the table, although its derivation has not been explained in this volume.

Table VI contains constants and their logarithms which will be of service in many computations. It will rarely be necessary to use more than seven of the nine decimals.

It may be noted that in these tables and throughout this book the logarithms of numbers less than unity are written with a negative characteristic; thus, log sin 0° 01′ is $\bar{2}.24903$, not 8.24903 as given in most tables. When using logarithmic tables the student should note that 8 and 9 mean $\bar{2}$ and $\bar{1}$ and should write them thus in his computations.

Table I. Mean Celestial Refraction.

Apparent Altitude.		Refraction.	
°	′	′	″
0	00	34	54.1
	10	32	49.2
	20	30	52.3
	30	29	03.5
	40	27	22.5
	50	25	49.8
1	00	24	24.6
	10	23	06.7
	20	21	55.6
	30	20	50.9
	40	19	51.9
	50	18	58.0
2	00	18	08.6
	10	17	23.0
	20	16	40.7
	30	16	00.9
	40	15	23.4
	50	14	47.8
3	00	14	14.6
	10	13	43.7
	20		15.0
	30	12	48.3
	40		23.7
	50		00.7
4	00	11	38.9
	10		18.3
	20	10	58.6
	30		39.6
	40		21.2
	50		03.3
5	00	9	46.5
	10		30.9
	20		16.0
	30		01.9
	40	8	48.4
	50		35.6
6	00	8	23.3
	10		11.6
	20		00.3
	30	7	49.5
	40		39.2
	50		29.2
7	00	7	19.7
	10		10.5
	20		01.7
	30	6	53.3
	40		45.1
	50		37.2

Apparent Altitude.		Refraction.	
°	′	′	″
8	00	6	29.6
	20		15.2
	40		01.8
9	00	5	49.3
	20		37.6
	40		26.5
10	00	5	16.2
	20		06.4
	40	4	57.2
11	00	4	48.5
	20		40.2
	40		32.4
12	00	4	25.0
	20		18.0
	40		11.3
13	00	4	04.9
	20	3	58.8
	40		53.0
14	00	3	47.4
	20		42.1
	40		37.0
15	00	3	32.1
	20		27.4
	40		22.9
16	00	3	18.6
	20		14.5
	40		10.5
17	00	3	06.6
	20		02.9
	40	2	59.3
18	00	2	55.8
	20		52.5
	40		49.3
19	00	2	46.1
	20		43.1
	40		40.2
20	00	2	37.3
	20		34.5
	40		31.9
21	00	2	29.3
	20		26.8
	40		24.3
22	00	2	21.9
	30		18.5

Apparent Altitude.		Refraction.	
°	′	′	″
23	00	2	15.2
	30		12.0
24	00	2	08.9
	30		06.0
25	00	2	03.2
	30		00.5
26	00	1	57.8
	30		55.3
27	00	1	52.8
	30		50.5
28	00	1	48.2
	30		46.0
29	00	1	43.8
	30		41.7
30	00	1	39.7
	30		37.7
31	00	1	35.8
	30		33.9
32	00	1	32.1
	30		30.3
33	00	1	28.7
	30		27.0
34	00	1	25.4
	30		23.8
35	00	1	22.3
	30		20.8
36	00	1	19.3
	30		17.8
37	00	1	16.5
	30		15.1
38	00	1	13.8
	30		12.4
39	00	1	11.2
	30		09.9
40	00	1	08.7
	30		07.5
41	00	1	06.3
	30		05.1
42	00	1	04.0
	30		02.9

Apparent Altitude.	Refraction.	
°	′	″
43	1	01.8
44	0	59.7
45		57.7
46		55.7
47		53.8
48		51.9
49		50.2
50	0	48.4
51		46.7
52		45.1
53		43.5
54		41.9
55		40.4
56		38.9
57		37.5
58		36.1
59		34.7
60	0	33.3
61		32.0
62		30.7
63		29.4
64		28.2
65		26.9
66		25.7
67		24.5
68		23.3
69		22.2
70	0	21.0
71		19.9
72		18.8
73		17.7
74		16.6
75		15.5
76		14.5
77		13.4
78		12.3
79		11.2
80	0	10.2
81		09.1
82		08.1
83		07.1
84		06.1
85		05.1
86		04.1
87		03.1
88		02.1
89		01.1

Table II. Lengths of Arcs of the Meridian.

Latitude.	One Degree.		One Minute.	One Second.	
	Miles.	Meters.	Meters.	Meters.	Feet.
0°	68.703	110 568	1842.81	30.713	100.766
5	.709	577	2.95	.716	.773
10	.725	602	3.37	.723	.797
15	.751	644	4.06	.734	.834
20	68.786	110 700	1844.98	30.750	100.886
21	.794	713	5.21	.737	.897
22	.802	726	5.44	.741	.910
23	.810	740	5.68	.761	.922
24	.819	754	5.91	.765	.935
25	68.829	110 769	1846.15	30.769	100.949
26	.838	785	6.41	.773	.963
27	.848	800	6.67	.778	.977
28	.858	816	6.94	.782	.992
29	.868	833	7.21	.787	101.007
30	68.879	110 850	1847.49	30.791	101.022
31	.889	867	7.78	.796	.038
32	.900	884	8.07	.801	.054
33	.911	902	8.37	.806	.070
34	.923	920	8.67	.811	.086
35	68.934	110 939	1848.98	30.816	101.103
36	.946	957	1849.29	.821	.120
37	.957	976	9.60	.827	.137
38	.969	995	9.92	.832	.155
39	.981	111 014	1850.24	.837	.172
40	68.993	111 034	1850.56	30.843	101.190
41	69.005	053	0.89	.848	.208
42	.017	073	1.22	.854	.225
43	.029	093	1.54	.859	.243
44	.042	112	1.87	.865	.261
45	69.054	111 132	1852.20	30.870	101.279
46	.067	152	2.53	.876	.297
47	.079	172	2.86	.881	.315
48	.091	191	3.19	.886	.333
49	.103	211	3.51	.892	.351
50	69.115	111 231	1853.84	30.897	101.369
51	.127	250	4.16	.903	.387
52	.139	269	4.49	.908	.404
53	.151	288	4.81	.913	.422
54	.163	307	5.12	.919	.439
55	69.175	111 326	1855.43	30.924	101.456
60	.230	416	5.69	.949	.538
65	.281	497	5.82	.971	.612
70	.324	567	5.95	.991	.676
75	69.360	111 624	1860.40	31.007	101.728
80	.386	666	1.10	.018	.766
85	.402	692	1.53	.026	.788
90	.407	701	1.68	.028	.797

Table III. Lengths of Arcs of Parallels.

Latitude.	One Degree.		One Minute.	One Second.	
	Miles.	Meters.	Meters.	Meters.	Feet.
0°	69.171	111 322	1855.36	30.923	101.452
5	68.911	101 901	1848.35	30.806	101.069
10	68.128	109 642	1827.36	30.456	99.921
15	66.830	107 553	1792.55	29.709	98.018
20	65.026	104 650	1744.16	29.069	95.372
21	64.606	103 973	1732.89	28.881	94.755
22	64.166	103 265	1721.08	28.685	94.110
23	63.706	102 525	1708.76	28.479	93.436
24	63.227	101 755	1695.91	28.265	92.733
25	62.729	100 953	1682.55	28.042	92.003
26	62.212	100 121	1668.68	27.811	91.244
27	61.676	99 258	1654.30	27.562	90.458
28	61.121	98 365	1639.41	27.322	89.644
29	60.548	97 442	1624.03	27.067	88.803
30	59.956	96 489	1608.16	26.803	87.935
31	59.345	95 507	1591.79	26.530	87.040
32	58.717	94 496	1574.94	26.249	86.118
33	58.071	93 456	1557.61	25.960	85.171
34	57.407	92 388	1539.80	25.663	84.197
35	56.726	91 291	1521.52	25.359	83.198
36	56.027	90 167	1502.78	25.046	82.173
37	55.311	89 015	1483.58	24.726	81.123
38	54.578	87 836	1463.93	24.395	80.048
39	53.829	86 630	1443.83	24.064	78.949
40	53.063	85 937	1432.28	23.871	77.826
41	52.281	84 138	1402.31	23.372	76.679
42	51.483	82 854	1380.90	23.015	75.508
43	50.669	81 544	1359.07	22.651	74.315
44	49.840	80 209	1336.82	22.280	73.098
45	48.995	78 850	1314.17	21.903	71.859
46	48.135	77 466	1291.11	21.518	70.599
47	47.261	76 059	1267.65	21.128	69.316
48	46.372	74 629	1243.81	20.730	68.012
49	45.469	73 175	1219.58	20.362	66.687
50	44.552	71 699	1194.65	19.911	65.342
51	43.621	70 201	1170.01	19.500	63.977
52	42.676	68 681	1144.68	19.078	62.592
53	41.719	67 140	1119.01	18.650	61.188
54	40.749	65 579	1092.98	18.216	59.765
55	39.766	63 997	1066.62	17.777	58.323
60	34.674	55 803	930.05	15.501	50.855
65	29.315	47 178	786.30	13.105	42.995
70	23.729	38 189	636.48	10.608	34.803
75	17.960	28 904	481.73	8.029	26.341
80	12.051	19 395	323.24	5.387	17.675
85	6.049	9 735	162.25	2.704	8.871
90	0	0	0	0	0

Table IV. Logarithms for Geodetic Computations.

IN METERS.

Latitude.	Radius of Curvature. Of Meridian, R_1.	Radius of Curvature. Of Prime Vertical Normal Section, R_2.	Radius of Curvature. Of Parallel, r.	Factor m for Spherical Excess.
0°	6.801 7538	6.804 7034	6.804 7034	$\bar{9}$.40 694
5	7873	7137	.803 0579	689
10	8868	7478	.798 0993	676
15	6.802 0492	8019	.789 7457	664
20	6.802 2698	6.804 8754	6.777 8612	$\bar{9}$.40 624
21	3204	8922	.775 0439	618
22	3729	9098	.772 0757	611
23	4274	9279	.768 9540	604
24	4838	9467	.765 6769	596
25	6.802 5418	6.804 9661	6.762 2418	$\bar{9}$.40 589
26	6016	9860	.758 6462	582
27	6633	6.805 0065	.754 8874	571
28	7264	0275	.750 9624	565
29	7918	0492	.746 8685	555
30	6.802 8572	6.805 0712	6.742 6018	$\bar{9}$.40 546
31	9246	0938	.738 1594	536
32	9933	1176	.733 5382	527
33	6.803 0631	1399	.728 7313	518
34	1341	1635	.723 7377	509
35	6.803 2062	6.805 1876	6.718 5611	$\bar{9}$.40 500
36	2791	2118	.713 1694	490
37	3528	2364	.707 5850	480
38	4273	2613	.701 7934	470
39	5025	2863	.695 7889	460
40	6.803 5782	6.805 3116	6.689 5656	$\bar{9}$.40 450
41	6545	3369	.683 1168	440
42	7311	3625	.676 4350	430
43	8080	3880	.669 5155	419
44	8851	4138	.662 3479	409
45	6.803 9623	6.805 4395	6.653 6575	$\bar{9}$.40 399
46	6.804 0395	4652	.647 2365	389
47	1166	4908	.639 2741	379
48	1935	5166	.631 0275	368
49	2705	5422	.622 4851	358
50	6.804 3466	6.805 5677	6.613 7252	$\bar{9}$.40 348
51	4224	5929	.604 4647	338
52	4976	6180	.594 9600	328
53	5723	6429	.585 1059	318
54	7195	6676	.574 8863	301
55	6.804 8626	6.805 6919	6.564 2832	$\bar{9}$.40 284
60	6.805 0693	8086	.504 7786	252
65	3857	9141	.431 8624	210
70	6591	6.806 0052	.340 0569	173
75	6.805 8809	6.806 0791	6.219 0753	$\bar{9}$.40 144
80	6.806 0443	1336	.045 8038	125
85	1444	1670	5.746 4630	108
90	1782	1782	— ∞	104

Table V. Logarithms for the *LMZ* Problem.

IN METERS.

Latitude.	*A'*	*B*	*C*	*D*	*E*
0°	$\bar{2}$.509 9613	$\bar{2}$.512 6713	− ∞	− ∞	$\overline{15}$.6124
5	7114	6378	$\overline{10}$.34885	$\bar{9}$.6275	.6223
10	6773	5383	.65308	.9220	.6511
15	6232	3759	.38460	$\bar{8}$.0871	.6970
20	$\bar{2}$.509 5497	$\bar{2}$.512 1553	$\overline{10}$.96732	$\bar{8}$.1964	$\overline{15}$.7574
21	5329	1047	.99036	.2138	.7712
22	5153	0522	$\bar{9}$.01252	.2302	.7852
23	4972	$\bar{2}$.511 9977	.03389	.2487	.7997
24	4784	9613	.05455	.2629	.8146
25	$\bar{2}$.509 4592	$\bar{2}$.511 8834	$\bar{9}$.07456	$\bar{8}$.2762	$\overline{15}$.8300
26	4391	8231	.09399	.2885	.8458
27	4186	7619	.11289	.3000	.8620
28	3976	6988	.13131	.3107	.8785
29	3760	6341	.14931	.3206	.8955
30	$\bar{2}$.509 3540	$\bar{2}$.511 5681	$\bar{9}$.16691	$\bar{8}$.3298	$\overline{15}$.9127
31	3315	5006	.18415	.3382	.9304
32	3086	4320	.20107	.3460	.9484
33	2853	3621	.21771	.3532	.9667
34	2617	2911	.23408	.3597	.9853
35	$\bar{2}$.509 2377	$\bar{2}$.511 2191	$\bar{9}$.25023	$\bar{8}$.3656	$\overline{14}$.0043
36	2134	1462	.26616	.3709	.0237
37	1888	0724	.28192	.3756	.0433
38	1639	$\bar{2}$.510 9979	.29752	.3797	.0633
39	1389	9227	.31298	.3833	.0836
40	$\bar{2}$.509 1136	$\bar{2}$.510 8469	$\bar{9}$.32833	$\bar{8}$.3863	$\overline{14}$.1043
41	0882	7707	.34357	.3888	.1253
42	0627	6941	.35874	.3907	.1467
43	0471	6172	.37385	.3921	.1684
44	0114	5401	.38893	.3929	.1905
45	2.508 9856	$\bar{2}$.510 4629	$\bar{9}$.40399	$\bar{8}$.3933	$\overline{14}$.2130
46	9599	3857	.41905	.3932	.2359
47	9342	3086	.43413	.3924	.2592
48	9065	2316	.44925	.3912	.2830
49	8830	1550	.46442	.3894	.3071
50	$\bar{2}$.508 8575	$\bar{2}$.510 0787	$\bar{9}$.47967	$\bar{8}$.3871	$\overline{14}$.3318
51	8323	0028	.49501	.3842	.3569
52	8072	$\bar{2}$.509 9275	.51047	.3808	.3826
53	7823	8529	.52607	.3768	.4088
54	7576	7790	.54182	.3722	.4355
55	$\bar{2}$.508 7333	$\bar{2}$.509 7059	$\bar{9}$.55776	$\bar{8}$.3671	$\overline{14}$.4629
60	6166	3559	.64108	.3320	.6102
65	5111	0394	.73342	.2790	.7802
70	4199	$\bar{2}$.508 7660	.84066	.1998	.9836
75	2.508 3460	$\bar{2}$.508 5442	$\bar{9}$.97338	$\bar{8}$.0909	$\overline{13}$.2410
80	2915	3808	$\bar{8}$.15493	$\bar{9}$.9262	.9986
85	2581	2807	.45913	.6319	$\overline{12}$.2038
90	2469	2469	− ∞	− ∞	+ ∞

TABLE VI. CONSTANTS AND THEIR LOGARITHMS.

Name. (Radius of circle or sphere = 1.)	Symbol.	Number.	Logarithm.
Area of circle	π	3 141 592 654	0.497 149 873
Circumference of circle	2π	6.283 185 307	0.798 179 868
Surface of sphere	4π	12.566 370 614	1.099 209 864
	$\frac{1}{6}\pi$	0.523 598 776	$\bar{1}$.718 998 622
Quadrant of circle	$\frac{1}{4}\pi$	0.785 398 163	$\bar{1}$.895 089 881
Area of semicircle	$\frac{1}{2}\pi$	1.570 796 327	0.196 119 877
Volume of sphere	$\frac{4}{3}\pi$	4.187 790 205	0.622 088 609
	π^2	9.869 604 401	0.994 299 745
	$\pi^{\frac{1}{2}}$	1.772 453 851	0.248 574 936
Degrees in a radian	$180/\pi$	57.295 779 513	1.758 122 632
Minutes in a radian	$10800/\pi$	3 437.746 771	3.536 273 883
Seconds in a radian	$648000/\pi$	206 264.806	5.314 425 133
	$1/\pi$	0.318 309 886	$\bar{1}$.502 850 127
	$1/\pi^{\frac{1}{2}}$	0.564 189 584	$\bar{1}$.751 425 064
	$1/\pi^2$	0.101 321 184	$\bar{1}$.005 700 255
Circumference/360	arc 1°	0.017 453 293	$\bar{2}$.241 877 368
	sin 1°	0.017 452 406	$\bar{2}$.241 855 318
Circumference/21600	arc 1′	0.000 290 888	$\bar{4}$.463 726 117
	sin 1′	0.000 290 888	$\bar{4}$.463 726 111
Circumference/1296000	arc 1″	0.000 004 848	$\bar{6}$.685 574 867
	sin 1″	0.000 004 848	$\bar{6}$.685 574 867
Base Naperian system of logs	e	2.718 281 828	0.434 294 482
Modulus common system of logs	M	0.434 294 482	$\bar{1}$.637 784 311
Naperian log of 10	$1/M$	2.302 585 093	0.362 215 689
	hr	0.476 936 3	$\bar{1}$.678 460 4
Probable error constant	$hr\sqrt{2}$	0.674 489 7	$\bar{1}$.828 975 4
Feet in one meter	m/ft.	3.280 833 3	0.515 984 1
Miles in one kilometer	km/mi.	0.621 369 9	$\bar{1}$.793 350 2

INDEX.